Structural Timber Design

Abdy Kermani   BSc, MSc, PhD, FIWSc
Napier University, Edinburgh

b

**Blackwell
Science**

© 1999 by
Blackwell Science Ltd
Editorial Offices:
Osney Mead, Oxford OX2 0EL
25 John Street, London WC1N 2BL
23 Ainslie Place, Edinburgh EH3 6AJ
350 Main Street, Malden
  MA 02148 5018, USA
54 University Street, Carlton
  Victoria 3053, Australia
10, rue Casimir Delavigne
  75006 Paris, France

Other Editorial Offices:

Blackwell Wissenschafts-Verlag GmbH
Kurfürstendamm 57
10707 Berlin, Germany

Blackwell Science KK
MG Kodenmacho Building
7–10 Kodenmacho Nihombashi
Chuo-ku, Tokyo 104, Japan

First published 1999

Set in 10/12.5 pt Times
by Aarontype Limited, Easton, Bristol
Printed and bound in Great Britain by
The University Press, Cambridge

The Blackwell Science logo is a trade mark of
Blackwell Science Ltd, registered at the United
Kingdom Trade Marks Registry

DISTRIBUTORS
Marston Book Services Ltd
PO Box 269
Abingdon
Oxon OX14 4YN
(*Orders*: Tel: 01235 465500
          Fax: 01235 465555)

USA
Blackwell Science, Inc.
Commerce Place
350 Main Street
Malden, MA 02148 5018
(*Orders*: Tel: 800 759 6102
                   781 388 8250
          Fax: 781 388 8255)

Canada
  Login Brothers Book Company
  324 Saulteaux Cresent
  Winnipeg, Manitoba R3J 3T2
  (*Orders*: Tel: 204 837 2987
            Fax: 204 837 3116)

Australia
  Blackwell Science Pty Ltd
  54 University Street
  Carlton, Victoria 3053
  (*Orders*: Tel: 03 9347 0300
            Fax: 03 9347 5001)

A catalogue record for this title is available
from the British Library

ISBN 0-632-05091-8

For further information on
Blackwell Science, visit our website:
www.blackwell-science.com

To Roman

Contents

Preface

The increasing recognition of timber as a structural material is reflected in the inclusion of timber design in many undergraduate courses. However, the majority of design textbooks for undergraduate engineering students neglect, to a large extent, the importance of timber as a structural and building material. As a consequence, relatively few textbooks provide information on the design of timber structures. *Structural Timber Design* is intended to address this issue by providing a step-by-step approach to the design of all the most commonly used timber elements and joints illustrated by detailed worked examples. This is an approach which is recognised to be beneficial in learning and preferred by most students.

The book has been written for undergraduate students on building, civil and structural engineering and architectural courses and will be an invaluable reference source and design aid for practising engineers and postgraduate engineering students. It provides a comprehensive source of information on practical timber design and encourages the use of computers to carry out design calculations.

Chapter 1 introduces the nature and inherent characteristics of timber such as defects, moisture content and slope of grain, and discusses the types of timber and factors that influence their structural characteristics. Chapter 2 includes a comprehensive review of the recently revised British Standard BS 5268 : Part 2 : 1996: *Structural Use of Timber*. The design philosophy of BS 5268 and its new approach to the strength class system and also the factors affecting timber strength are explained.

Chapter 3 gives an overview of Mathcad®, a computer software programme used to carry out mathematical calculations, and details its simplicity and the advantages that it provides when used for design calculations. The aim is to encourage readers to use computing as a tool to increase their under-standing of how design solutions vary in response to a change in one of the variables and how alternative design options can be obtained easily and effortlessly. The design of basic elements is explained and illustrated in Chapters 4 and 5, whilst the design of more specialised elements such as glued laminated straight and curved beams and columns, ply-webbed beams and built-up columns is illustrated in Chapters 6, 7 and 8 using numerous worked examples.

In Chapter 9 the design of timber connections is detailed. The new approach adopted by the revised BS 5268 : Part 2 in 1996, i.e. the Eurocode 5 approach for the design of timber joints, is described. The chapter includes a comprehensive coverage of the design requirements for nailed, screwed, bolted and dowelled joints, and the design of connectored joints such as toothed-plates, split-rings and shear-plates and glued connections is also detailed. Several step-by-step worked examples are provided to illustrate the design methods in this chapter.

Chapter 10 provides a comprehensive review of the proposed European code for timber, Eurocode 5: *Design of Timber Structures*. The limit states design philosophy of EC5 is explained and the relevant differences with the design methodology of BS 5268 are highlighted and discussed. This chapter also provides comprehensive coverage of EC5 requirements for the design of flexural and axially loaded members and dowel-type connections such as nailed, screwed, bolted and dowelled joints. Again, step-by-step worked examples are provided to illustrate the design methods in the chapter.

All design examples given in this book are produced in the form of worksheet files and are available from the author on $3\frac{1}{2}''$ disks to run under Mathcad computer software Version 6, or higher, in either one of its editions: (Student, Standard, Plus or Professional). Details are given at the end of the book. The examples are fully self-explanatory and well annotated and the author is confident that the readers whether students, course instructors, or practising design engineers will find them extremely useful to produce design solutions or prepare course handouts. In particular, the worksheets will allow design engineers to arrive at the most suitable/economic solution(s) very quickly.

Extracts from British Standards are reproduced with the permission of BSI under licence no. PD\1998 0823. Complete editions of the standards can be obtained by post from BSI Customer Services, 389 Chiswick High Road, London W4 4AL.

The cover illustration was kindly supplied by MiTek Industries Ltd.

Chapter 1
Timber as a Structural Material

1.1 Introduction

Timber has always been one of the more plentiful natural resources available and consequently is one of the oldest known materials used in construction. It is a material that is used for a variety of structural forms such as beams, columns, trusses, girders and is also used in building systems such as piles, deck members, railway foundations and for temporary forms in concrete.

Timber structures can be highly durable when properly treated and built. Examples of this are seen in many historic buildings all around the world. Timber possesses excellent insulating properties, good fire resistance, light weight and aesthetic appeal. A great deal of research carried out since the early part of this century has provided us with comprehensive information on structural properties of timber and timber products[1].

A knowledge of engineering materials is essential for engineering design. Timber is a traditional building material and over the years considerable knowledge has been gained on its important material properties and their effects on structural design and service behaviour. Many failures in timber buildings in the past have shown us the safe methods of construction, connection details and design limitations.

This chapter provides a brief description of the engineering properties of timber that are of interest to design engineers or architects. But it should be kept in mind that, unlike some structural materials such as steel or concrete, the properties of timber are very sensitive to environmental conditions. For example, timber is very sensitive to moisture content, which has a direct effect on the strength and stiffness, swelling or shrinkage of timber. A proper understanding of the physical characteristics of wood aids the building of safe timber structures[1].

1.2 The structure of timber [2]

Mature trees of whatever type are the source of structural timber and it is important that users of timber should have a knowledge of the nature and growth patterns of trees in order to understand its behaviour under a variety of circumstances. Basically, a tree has three subsystems: *roots*, *trunk* and *crown*. Each subsystem has a role to play in the growth pattern of the tree.

(1) *Roots*, by spreading through the soil as well as acting as a foundation, enable the growing tree to withstand wind forces. They absorb moisture containing minerals from the soil and transfer it via the trunk to the crown.
(2) *Trunk* provides rigidity, mechanical strength and height to maintain the crown. Also transports moisture and minerals up to the crown and sap down from the crown.
(3) *Crown* provides as large as possible a catchment area covered by leaves. These produce chemical reactions that form sugar and cellulose which cause the growth of the tree.

As engineers we are mainly concerned with the trunk of the tree. Consider a cross-section of a trunk as shown in Fig. 1.1.

Wood, in general, is composed of long thin tubular cells. The cell walls are made up of cellulose and the cells are bound together by a substance known as lignin. Most cells are oriented in the direction of the axis of the trunk, except for cells known as *rays* which run radially across the trunk. Rays are present in all trees but are more pronounced in some species, such as oak. In temperate countries, a tree produces a new layer of wood just under the bark in the early part of every growing season. This growth ceases at the end of the growing season or during winter months. This process results in clearly visible concentric rings known as *annular rings*, *annual rings* or *growth rings*. In tropical countries where trees grow throughout the year, a tree produces wood cells that are essentially uniform. The age of a tree may be determined by counting its growth rings[1].

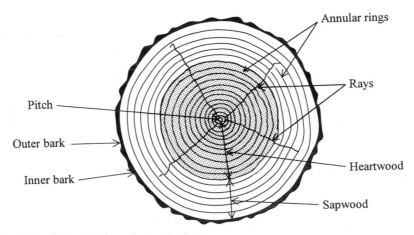

Fig. 1.1   Cross-section of a trunk of a tree.

The annular band of cross-section nearest to the bark is called *sapwood*. The central core of the wood which is inside the sapwood is *heartwood*. The sapwood is lighter in colour compared to heartwood and is 25–170 mm wide, depending on the species. It acts as a medium of transportation for sap from the roots to the leaves, while the heartwood functions mainly to give mechanical support or stiffness to the trunk. In general, the moisture content, strength and weights of the two are nearly equal. Sapwood has a lower natural resistance to attacks by fungi and insects and accepts preservatives more easily than heartwood[1].

In many trees, each annular ring can be subdivided into two layers: an inner layer made up of relatively large cavities called *springwood*, and an outer layer of thick walls and small cavities called *summerwood*. Since summerwood is relatively heavy, the amount of summerwood in any section is a measure of the density of the wood.

1.3 Defects in timber[2]

Owing to the fact that wood is a material which is naturally occurring, there are many defects which are introduced during the growing period and during the conversion and seasoning process. Any of these defects can cause trouble in timber in use either by reducing its strength or impairing its appearance.

Defects may be classified as: natural defects, chemical defects, conversion defects and seasoning defects.

1.3.1 *Natural defects*

These occur during the growing period. Examples of natural defects are illustrated in Fig. 1.2(a). These may include:

- *Cracks and fissures*. They may occur in various parts of the tree and may even indicate the presence of decay or the beginnings of decay.
- *Knots*. These are common features of the structure of wood. A knot is a portion of a branch embedded by the natural growth of the tree, normally originating at the centre of the trunk or a branch.
- *Grain defects*. Wood grain refers to the general direction of the arrangement of fibres in wood. Grain defects can occur in the form of twisted-grain, cross-grain, flat-grain and spiral-grain, all of which can induce subsequent problems of distortion in use.
- *Fungal decay*. This may occur in growing mature timber or even in recently converted timber, and in general it is good practice to reject such timber.
- *Annual ring width*. This can be critical in respect of strength in that excess width of such rings can reduce the density of the timber.

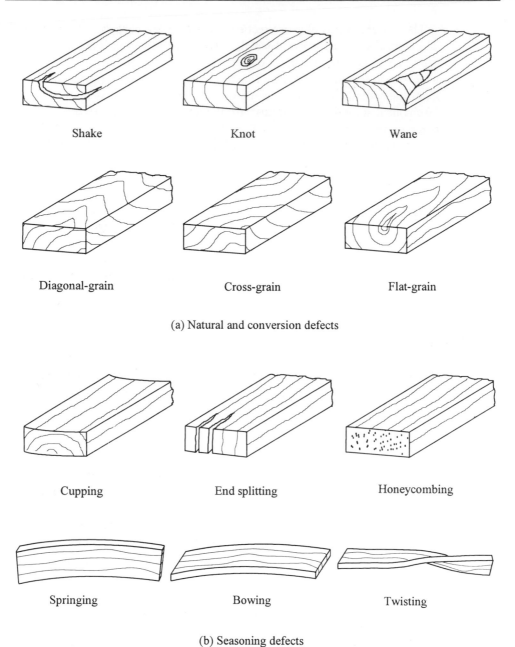

(a) Natural and conversion defects

(b) Seasoning defects

Fig. 1.2   Defects in timber.

1.3.2   *Chemical defects*

These may occur in particular instances when timber is used in unsuitable positions or in association with other materials. Timbers such as oak and western red cedar contain tannic acid and other chemicals which corrode metals. Gums and resins can inhibit the working properties of timber and interfere with the ability to take adhesives.

1.3.3 *Conversion defects*

These are due basically to unsound practice in the use of milling techniques or to undue economy in attempting to use every possible piece of timber converted from the trunk. A wane is a good example of a conversion defect.

1.3.4 *Seasoning defects*

Seasoning defects are directly related to the movement that occurs in timber due to changes in moisture content. Excessive or uneven drying, exposure to wind and rain, and poor stacking and spacing during seasoning can all produce defects or distortions in timber. Examples of seasoning defects are illustrated in Fig. 1.2(b). All such defects have an effect on structural strength as well as on fixing, stability, durability and finished appearance.

1.4 Types of timber

Trees and commercial timbers are divided into two groups: *softwoods* and *hardwoods*. This terminology has no direct bearing on the actual softness or hardness of the wood.

1.4.1 *Softwoods*

Softwoods are generally evergreen with needle-like leaves comprising single cells called *tracheids*, which are like straws in plan, and they fulfil the functions of conduction and support. Rays, present in softwoods, run in a radial direction perpendicular to the growth rings. Their function is to store food and allow the convection of liquids to where they are needed.

Softwood characteristics

- Quick growth rate; trees can be felled after 30 years, resulting in low density timber with relatively low strength.
- Generally poor durability qualities, unless treated with preservatives.
- Due to speed of felling, they are readily available and comparatively cheap.

1.4.2 *Hardwoods*

Hardwoods are generally broad-leaved (deciduous) trees that lose their leaves at the end of each growing season. The cell structure of hardwoods is more complex than that of softwoods, with thick walled cells, called *fibres*, providing the structural support and thin walled cells, called *vessels*, providing the medium for food conduction. Due to the necessity to grow new leaves every year the demand for sap is high and in some instances larger vessels may be formed in the springwood – these are referred to as *rig porous* woods. When there is no definite growing period the pores tend to be more evenly distributed, resulting in *diffuse porous* woods.

Hardwood characteristics

- Hardwoods grow at a slower rate than softwoods. This generally results in a timber of high density and strength which takes time to mature – over 100 years in some instances.
- There is less dependency on preservatives for durability qualities.
- Due to time taken to mature and the transportation costs of hardwoods, as most are tropical, they tend to be expensive in comparison to softwoods.

1.5 Physical properties of timber[3]

Due to the fact that timber is such a variable material, its strength is dependent on many factors which can act independently or in conjunction with others, adversely affecting the strength and the workability of the timber. Among many physical properties that influence the strength characteristics of timber, the following may be considered the most important ones.

1.5.1 *Moisture content*

The strength of timber is dependent on its moisture content, as is the resistance to decay. Most timber in the UK is air-dried to a moisture content of between 17% and 23% which is generally below fibre saturation point at which the cell walls are still saturated but moisture is removed from the cells. Any further reduction will result in shrinkage[4]. Figure 1.3 highlights the general relationship between strength and/or stiffness characteristics of timber and its moisture content. The figure shows that there is an almost linear loss in strength and stiffness as moisture increases to about 30%, corresponding to fibre saturation point. Further increases in moisture content have no influence on either strength or stiffness. It should be noted that, although for most mechanical properties the pattern of change in strength and stiffness characteristics with respect to change in moisture content is similar, the magnitude of change is different from one property to another. It is also to be noted that as the moisture content decreases shrinkage increases. Timber is described as being hygroscopic which means

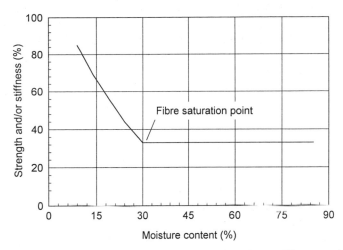

Fig. 1.3 General relationship between strength and/or stiffness and moisture content.

that it attempts to attain an equilibrium moisture content with its surrounding environment, resulting in a variable moisture content. This should always be considered when using timber, particularly softwoods which are more susceptible to shrinkage than hardwoods.

1.5.2 *Density*

Density is the best single indicator of the properties of a timber and is a major factor determining its strength. Specific gravity or relative density is a measure of timber's solid substance. It is generally expressed as the ratio of the oven-dry weight to the weight of an equal volume of water. Since water volume varies with the moisture content of the timber, the specific gravity of timber is expressed at a certain moisture content. Basic specific gravity of commercial timber ranges from 0.29 to 0.81, most falling between 0.35 and 0.60.

1.5.3 *Slope of grain*

Grain is the longitudinal direction of the main elements of timber, these main elements being fibres or tracheids, and vessels in the case of hardwoods. In many instances the angle of the grain in a cut section of timber is not parallel to the longitudinal axis. It is possible that this variation is due to poor cutting of the timber, but more often than not the deviation in grain angle is due to irregular growth of the tree. This effect is of lesser consequence when timber is axially loaded, but leads to a significant drop in bending resistance. The angle of the microfibrils within the timber also affects the strength of the timber, as with the effects of the grain, if the angle of deviation increases the strength decreases.

1.5.4 *Timber defects*

As described earlier, defects in timber, whether natural or caused during conversion or seasoning, will have an effect on structural strength as well as on fixing, stability, durability and finished appearance of timber.

1.6   References

1. Somayaji, S. (1990) *Structural Wood Design.* West Publishing Company, St. Paul, U.S.A.
2. Illston, J.M., Dinwoodie, J.M. and Smith, A.A. (1979) *Concrete, Timber and Metals – The Nature and Behaviour of Structural Materials.* Van Nostrand Reinhold International, London.
3. Illston, J.M. (1994) *Construction Materials – Their Nature and Behaviour.* E. & F.N. Spon, London.
4. Carmichael, E.N. (1984) *Timber Engineering.* E. & F.N. Spon, London.

Chapter 2
Introduction to BS 5268 : Part 2 : 1996

2.1 Introduction

Strength capability of timber is difficult to assess as we have no control over its quality and growth. The strength of timber is a function of several parameters including the moisture content, density, duration of the applied load, size of members and presence of various strength-reducing characteristics such as slope of grain, knots, fissures and wane. To overcome this difficulty, the stress grading method of strength classification has been devised[1].

Guidance on the use of timber in building and civil engineering structures is given in BS 5268: *Structural use of timber*. This was originally divided into seven parts:

Part 1: Limit state design, materials and workmanship.
Part 2: Code of practice for permissible stress design, materials and workmanship.
Part 3: Code of practice for trussed rafter roofs.
Part 4: Fire resistance of timber structures.
Part 5: Preservation treatments for constructional timber.
Part 6: Code of practice for timber framed walls.
Part 7: Recommendations for the calculation basis for span tables.

Part 1 of BS 5268 was never completed and, with the introduction of Eurocode 5: DD ENV 1995-1-1: *Design of timber structures*, the development of this part was completely abandoned.

Part 2 of BS 5268, on which the design of structural timber is based, was originally published as CP 112 in 1952 and revised later in 1967 and, with extensive amendment, in 1971. The 'basic stresses' introduced in CP 112 were determined from carrying out short-term loading tests on small timber specimens free from all defects. The data was used to estimate the minimum strength which was taken as the value below which not more than 1% of the test results fell. These strengths were multiplied by a reduction factor to give basic stresses. The reduction factor made an allowance for the reduction in strength due to duration for loading, size of specimen and other effects normally associated with a safety factor, such as accidental overloading, simplifying assumptions made during design and design inaccuracy, together with poor workmanship. Basic stress was defined as the stress that could be permanently sustained by timber free from any strength-reducing characteristics[1].

Since 1967 there have been continuing and significant changes affecting the structural use of timber. Research studies in the UK and other countries had shown the need for a review of the stress values and modification factors given in the original code.

With the introduction of BS 5268 in 1984 the concept of 'basic stresses' was largely abandoned and the new approach for assessing the strength of timber moved somewhat in line with 'limit states' design philosophy. In 1996, Part 2 of BS 5268 was revised with a clear aim to bring this code as close as possible to, and to run in parallel with, Eurocode 5: DD ENV 1995-1-1: *Design of timber structures, Part 1.1 General rules and rules for buildings*. The overall aim has been to incorporate material specifications and design approaches from Eurocode 5, while maintaining a permissible stress code with which designers, accustomed to BS 5268, will feel familiar and be able to use without difficulty. The first step in this process involves strength grading of timber sections. There are two European standards which relate to strength grading:

> BS EN 518:1995 *Structural timber. Grading. Requirements for visual strength grading standards.*
> BS EN 519:1995 *Structural timber. Grading. Requirements for machine strength graded timber and grading machines.*

Guidance for stress grading of the two types of timber, namely softwoods and hardwoods, are given in the following British Standards:

> BS 4978:1996 *Specification for softwoods graded for structural use.*
> BS 5756:1997 *Specification for tropical hardwoods graded for structural use.*

The current revised versions of these standards conform with the requirements of BS EN 518:1995.

2.2 Design philosophy

The structural design of timber members is related to Part 2 of BS 5268, and is based on *permissible stress* design philosophy in which design stresses are derived on a statistical basis and deformations are also limited.

Elastic theory is used to analyse structures under various loading conditions to give the worst design case. Then timber sections are chosen so that the permissible stresses are not exceeded at any point of the structure.

Permissible stresses are calculated by multiplying the 'grade stresses', given in Tables 7 to 12a of BS 5268 : Part 2, by the appropriate modification factors, *K-factors*, to allow for the effects of parameters such as load duration, moisture content, load sharing, section size, etc. Applied stresses which are derived from the service loads should be less than or equal to the permissible stresses. A summary of the K-factors used for the calculation of permissible stresses is given in Table 2.1. Owing to changes made to BS 5268 : Part 2 in 1996, some K-factors which were used in the previous editions, such as K_1, K_{10}, etc., have been withdrawn.

The permissible stress design philosophy, as in BS 5268 : Part 2, is different from the limit states design philosophy of Eurocode 5 which has two basic requirements. The first is *ultimate limit states* (i.e. *safety*) which is usually

Table 2.1   Summary of K-factors used for calculation of permissible stresses

K-factor	Description or application	BS 5268 : Part 2 : 1996
K_2	Timber grade stresses and moduli for service class 3	Table 13
K_3	Duration of loading	Table 14
K_4	Bearing stress	Table 15
K_5	Shear at notched ends	Clause 2.10.4
K_6	Form factor: bending stress for non-rectangular sections	Clause 2.10.5
K_7	Depth factor: bending stress for beams other than 300 mm deep	Clause 2.10.6
K_8	Load sharing systems	Clause 2.9
K_9	To modify E_{min} for deflection in trimmer beams and lintels	Table 17
K_{12}	Slenderness in compression members	Table 19/Annex B
K_{13}	Effective length of spaced columns	Table 20
K_{14}	Width factor for tension members	Clause 2.12.2
K_{15-20}	Single grade glued laminated members and horizontally laminated beams	Table 21
K_{27-29}	Vertically glued laminated members	Table 22
K_{30-32}	Individually designed glued end joints in horizontally glued laminated members	Table 23
K_{33-34}	Curved glued laminated beams	Clause 3.5.3
K_{35}	Stress factor in pitched cambered softwood beams	Clause 3.5.4.2
K_{36}	Plywood grade stresses for duration of loading and service classes	Table 33
K_{37}	Stress concentration factor for ply-webbed beams	Clause 4.6
K_{38-41}	For tempered hardboards	Section 5
K_{ser}	Fastener slip moduli	Table 52
K_{43-50}	Nailed joints	Clause 6.4
K_{52-54}	Screwed joints	Clause 6.5
$K_{56,\,57}$	Bolted and dowelled joints	Clause 6.6
$K_{S,C,58-61}$	Toothed-plate connector joints	Clause 6.7
$K_{S,C,D,62-65}$	Split-ring connector joints	Clause 6.8
$K_{S,C,D,66-69}$	Shear-plate connector joints	Clause 6.9
K_{70}	Glued joints	Clause 6.10

expressed in terms of load-carrying capacity and is achieved by factoring-up of load values and factoring-down of material strength properties by partial safety factors that reflect the reliability of the values that they modify. The second is *serviceability limit states* (i.e. *deformation and vibration limits*) which refers to the ability of a structural system and its elements to perform satisfactorily in normal use.

It is important to note that in permissible stress design philosophy partial safety factors (i.e. modification factors) are applied only to material properties, i.e. for the calculation of permissible stresses, and not to the loading.

2.3   Stress grading of timber

Once timber has been seasoned it is stress graded; this grading will determine the strength class of the timber to satisfy the design requirements of BS 5268 : Part 2. Strength grading takes into account defects within the timber such as slope of grain, existence and extent of knots and fissures, etc.

All timber used for structural work needs to be strength graded by either visual inspection or by an approved strength grading machine. Clause 2.5 of BS 5268 : Part 2 deals with strength grading of timber.

2.3.1   *Visual grading*

Visual grading is a manual process carried out by an approved grader. The grader examines each piece of timber to check the size and frequency of specific physical characteristics or defects, e.g. knots, slope of grains, rate of growth, wane, resin pockets and distortion, etc.

The required specifications are given in BS 4978 and BS 5756 to determine if a piece of timber is accepted into one of the two visual stress grades or rejected. These are General Structural (GS) and Special Structural (SS) grades. Table 2 of BS 5268 : Part 2 (reproduced here as Table 2.2) refers to main softwood combinations of species visually graded in accordance with BS 4978.

2.3.2   *Machine grading*

Machine grading of timber sections is carried out on the principle that strength is related to stiffness. The machine exerts pressure and bending is induced at increments along timber length. The resulting deflection is then automatically measured and compared with pre-programmed criteria, which leads to the grading of timber section. BS 5268 : Part 2, Clause 2.5 specifies that machine graded timber, other than that carried out by North American Export Standard for Machine Stress-rated Lumber (e.g. 1450f-1.3E), should meet the requirements of BS EN 519. To this effect timber is graded directly to the strength class boundaries and marked accordingly.

In general less material is rejected if it is machine graded, however timber is also visually inspected during machine grading to ensure major defects do not exist.

2.4   Strength classes

The concept of grouping timber into strength classes was introduced into the UK with BS 5268 : Part 2 in 1984. Strength classes offer a number of advantages both to the designer and the supplier of timber. The designer can undertake his design without the need to check on the availability and price of a large number of species and grades which he might use. Suppliers can supply any of the species/grade combinations that meet the strength class called for in a specification. The concept also allows new species to be introduced onto the market without affecting existing specifications for timber.

The latest strength classes used in the current version of BS 5268 : Part 2 : 1996 relate to the European strength classes which are defined in BS EN 338 : 1995 *Structural timber. Strength classes.* There are a total of 16 strength classes, C14 to C40 for softwoods and D30 to D70 for hardwoods as given in Table 7 of BS 5268 : Part 2 : 1996 (reproduced here as Table 2.3). The number in each strength class refers to its 'characteristic bending strength' value, for example, C40 timber has a characteristic bending strength of $40 \, \text{N/mm}^2$. It is to be noted that characteristic strength values are considerably larger than the grade stress values used in BS 5268 : Part 2, as they do not include effects of long-term loading and safety factors.

Softwood grading: Softwoods which satisfy the requirements for strength classes given in BS EN 338 when graded in accordance with BS 4978 and American timber standards NLGA and NGRDL are given in Tables 2, 3, 4 and 5 of BS 5268 : Part 2. The new strength classes for softwoods are C14, C16, C18, C22, C24, TR26, C27, C30, C35 and C40. However it is likely that the old strength class system (i.e. SC1 to SC9) may be encountered for some time. A comparison of the lowest of the new strength class (C classes) against the most common old SC classes can be made: SC3 compares with C16, SC4 with C24, and SC5 with C27. TR26 timber, which is commonly used for axially loaded members (i.e. trussed rafters), is equivalent to the superseded M75 European redwood/whitewood.

Hardwood grading: Tropical hardwoods which satisfy the requirements for strength classes given in BS EN 338 when graded to HS grade in accordance with BS 5756 are given in Table 6 of BS 5268 : Part 2 : 1996. The strength classes for tropical hardwoods are D30, D35, D40, D50, D60 and D70.

Grade stresses: Grade stresses and moduli of elasticity for service classes 1 and 2 (described in Section 2.5.2) are given in Table 7 of BS 5268 : Part 2 for

Table 2.2 Softwood combinations of species and visual grades which satisfy the requirements for various strength classes. Timber graded in accordance with BS 4978 (Table 2, BS 5268 : Part 2)

Timber species	Strength classes						
	C14	C16	C18	C22	C24	C27	C30
Imported:							
Parana pine		GS			SS		
Caribbean pitch pine			GS			SS	
Redwood		GS			SS		
Whitewood		GS			SS		
Western red cedar	GS		SS				
Douglas fir-larch (Canada and USA)		GS			SS		
Hem-fir (Canada and USA)		GS			SS		
Spruce-pine-fir (Canada and USA)		GS			SS		
Sitka spruce (Canada)	GS		SS				
Western white woods (USA)	GS		SS				
Southern pine (USA)			GS		SS		
British grown:							
Douglas fir	GS		SS				
Larch		GS			SS		
British pine	GS			SS			
British spruce	GS		SS				

16 strength classes, and in Tables 8 to 12a for individual softwood and hardwood species and grades. Table 7 is reproduced here as Table 2.3.

2.5 Design considerations (factors affecting timber strength)

As mentioned previously, there are several factors which influence timber strength and hence they should be considered in the analysis–design process of all structural timber members, assemblies and frameworks. The main design criteria recommended by BS 5268 : Part 2, Clause 1.6 for consideration are listed below.

2.5.1 *Loading*

For the purpose of design, loading should be in accordance with BS 6399 : Parts 1, 2, and 3[2] and CP 3: Chapter V : Part 2[3] or other relevant standards, where applicable.

2.5.2 *Service classes*

Due to the effects of moisture content on mechanical properties of timber, the permissible property values should be those corresponding to one of the

Table 2.3 Grade stresses and moduli of elasticity for various strength classes: for service classes 1 and 2 (Table 7, BS 5268: Part 2)

Strength class[a]	Building ∥ to grain, $\sigma_{m,g,\parallel}$ (N/mm²)	Tension ∥ to grain, $\sigma_{t,g,\parallel}$ (N/mm²)	Compression ∥ to grain, $\sigma_{c,g,\parallel}$ (N/mm²)	Compression ⊥ to grain,[b] $\sigma_{c,g,\perp}$ (N/mm²)	Shear ∥ to grain, $\tau_{g,\parallel}$ (N/mm²)	Modulus of elasticity		Density[c]	
						E_{mean} (N/mm²)	E_{min}	ρ_k (kg/m³)	ρ_{mean}
C14	4.1	2.5	5.2	2.1	1.6	0.60	6 800	4 600	290	350
C16	5.3	3.2	6.8	2.2	1.7	0.67	8 800	5 800	310	370
C18	5.8	3.5	7.1	2.2	1.7	0.67	9 100	6 000	320	380
C22	6.8	4.1	7.5	2.3	1.7	0.71	9 700	6 500	340	410
C24	7.5	4.5	7.9	2.4	1.9	0.71	10 800	7 200	350	420
TR26[d]	10.0	6.0	8.2	2.5	2.0	1.10	11 000	7 400	370	450
C27	10.0	6.0	8.2	2.5	2.0	1.10	12 300	8 200	370	450
C30	11.0	6.6	8.6	2.7	2.2	1.20	12 300	8 200	380	460
C35	12.0	7.2	8.7	2.9	2.4	1.30	13 400	9 000	400	480
C40	13.0	7.8	8.7	3.0	2.6	1.40	14 500	10 000	420	500
D30	9.0	5.4	8.1	2.8	2.2	1.40	9 500	6 000	530	640
D35	11.0	6.6	8.6	3.4	2.6	1.70	10 000	6 500	560	670
D40	12.5	7.5	12.6	3.9	3.0	2.00	10 800	7 500	590	700
D50	16.0	9.6	15.2	4.5	3.5	2.20	15 000	12 600	650	780
D60	18.0	10.8	18.0	5.2	4.0	2.40	18 500	15 600	700	840
D70	23.0	13.8	23.0	6.0	4.6	2.60	21 000	18 000	900	1080

[a] Strength classes C14 to C40 and TR26 are for softwoods, D30 to D70 are for hardwoods
[b] If wane is allowed, the lower value should be used
[c] Characteristic density values are used only when designing joints. For calculation of dead load, the average density should be used
[d] The strength class TR26 is essentially for the manufacture of trussed rafters but may be used for other applications with the grade stresses and moduli given above. When used with the provisos given in BS 5268: Part 3 the grade stresses are similar to the former M75 redwood/whitwood so timber and trussed rafter designs to this M75 grade/species combination are interchangeable with timber and trussed rafter designs using the TR26 strength class.

Table 2.4 Modification factor K_2 for obtaining stresses and moduli applicable to service class 3 (Table 13, BS 5268 : Part 2)

Property	K_2
Bending parallel to grain	0.8
Tension parallel to grain	0.8
Compression parallel to grain	0.6
Compression perpendicular to grain	0.6
Shear parallel to grain	0.9
Mean and minimum modulus of elasticity	0.8

three service classes described in Clause 1.6.4 and given in Table 1 of BS 5268 : Part 2 : 1996. These are summarised below:

(1) *Service class 1* refers to timber used internally in a continuously heated building. The average moisture content likely to be attained in service condition is 12%.
(2) *Service class 2* refers to timber used in a covered building. The average moisture content likely to be attained in service condition if building is generally heated is 15%, and if unheated, 18%.
(3) *Service class 3* refers to timber used externally and fully exposed. The average moisture content likely to be attained in service condition is over 20%.

Grade stress and elastic moduli values given in Tables 7 to 12a of BS 5268 : Part 2 apply to various strength classes and timber species in service classes 1 and 2. For service class 3 condition they should be multiplied by the modification factor K_2 from Table 13 of the code (reproduced here as Table 2.4).

2.5.3 *Moisture content*

As moisture content affects the structural properties of timber significantly, BS 5268 : Part 2 : 1996 recommends that in order to reduce movement and creep under load the moisture content of timber and wood-based panels when installed should be close to that likely to be attained in service.

2.5.4 *Duration of loading*

Duration of load affects timber strength and therefore the permissible stresses. The grade stresses (Tables 7 to 12a) and the joint strengths given in BS 5268 : Part 2 are applicable to long-term loading. Because timber and wood-based materials can sustain a much greater load for a short period

Table 2.5 Modification factor K_3 for duration of loading (Table 14, BS 5268 : Part 2)

Duration of loading	K_3
Long-term: i.e. dead + permanent imposed[a]	1.00
Medium-term: i.e. dead + temporary imposed + snow	1.25
Short-term: i.e. dead + imposed + wind,[b] dead + imposed + snow + wind[b]	1.50
Very short-term: i.e. dead + imposed + wind (gust)[c]	1.75

[a] For uniformly distributed imposed floor loads $K_3 = 1$ except for type 2 and type 3 buildings (see Table 5 of BS 6399 : Part 1 : 1984[2]) where, for corridors, hallways, landings and stairways only, K_3 may be assumed to be 1.5.
[b] For wind, short-term category applies to class C (15 s gust) as defined in CP 3 : Chapter V : Part 2[3] or, where the largest diagonal dimension of the loaded area a , as defined in BS 6399 : Part 2,[2] exceeds 50 m.
[c] For wind, very short-term category applies to classes A and B (3 s or 5 s gust) as defined in CP 3 : Chapter V : Part 2[3] or, where the largest diagonal dimension of the loaded area a, as defined in BS 6399 : Part 2,[2] does not exceeds 50 m.

(a few minutes) than for a long period (several years), the grade stresses and the joint loads may be increased for other conditions of loading by the modification factors given in the appropriate sections of BS 5268 : Part 2.

Table 14 of BS 5268 : Part 2 (reproduced here as Table 2.5) gives the modification factor K_3 by which all grade stresses (excluding moduli of elasticity and shear moduli) should be multiplied for various durations of loading.

2.5.5 *Section size*

The bending, tension and compression and moduli of elasticity given in Part 2 of BS 5268 are applicable to materials 300 mm deep (or wide, for tension). Because these properties of timber are dependent on section size and size related grade effects, the grade stresses should be modified for section sizes other than 300 mm deep by the modification factors specified in the appropriate sections of the code.

In general, it is possible to design timber structures using any size of timber. However, since the specific use is normally not known at the time of conversion, sawmills tend to produce a range of standard sizes known as 'customary' sizes. Specifying such customary sizes will often result in greater availability and savings in cost[4].

The customary lengths and sizes produced by sawmills in the UK, normally available from stock, are given in Tables NA.1 to NA.4 of the

National Annex to BS EN 336 : 1995 which uses target sizes as the basis for the standard. Further information and details of the customary lengths and sizes are given in Appendix A.

2.5.6 *Load-sharing systems*

The grade stresses given in Part 2 of BS 5268 are applicable to individual pieces of structural timber. Where a number of pieces of timber (in general four or more) at a maximum spacing of 610 mm centre to centre act together to support a common load, then the grade stresses can be modified (increased) in accordance with the appropriate sections of the code.

In a load-sharing system such as rafters, joists, trusses or wall studs spaced at a maximum of 610 mm centre to centre, and which has adequate provision for the lateral distribution of loads by means of purlins, binders, boarding, battens, etc., the appropriate grade stresses can be multiplied by the load-sharing modification factor K_8 which has a value of 1.1. In addition, BS 5268 : Part 2 recommends that the mean modulus of elasticity should be used to calculate deflections and displacements induced by static loading conditions.

Therefore in a load-sharing system:

> Modification factor $K_8 = 1.1$
> Modulus of elasticity $E = E_{mean}$

It is to be noted that special provisions are provided in BS 5268 : Part 2 for built-up beams, trimmer joists and lintels, and laminated beams; these are given in Clauses 2.10.10, 2.10.11 and Section 3 of the code. It is also important to note that the provisions for load-sharing systems do not extend to the calculation of modification factor K_{12} for load-sharing columns.

2.5.7 *Additional properties*

BS 5268 : Part 2 recommends that in the absence of test data, the following grade stress and moduli of elasticity values may be used:

tension perpendicular to grain, $\sigma_{t,g,\perp}$	$= \frac{1}{3} \times$ shear stress parallel to grain, $\tau_{g,\parallel}$
torsional shear, $\tau_{torsion}$	$= \frac{1}{3} \times$ shear stress parallel to grain, $\tau_{g,\parallel}$
rolling shear, τ_r	$= \frac{1}{3} \times$ shear stress parallel to grain, $\tau_{g,\parallel}$
modulus of elasticity $\perp$ to grain, $E_\perp$	$= \frac{1}{20} \times E_{mean\ or\ min}$
shear modulus, G	$= \frac{1}{16} \times E_{mean\ or\ min}$
permissible compressive stress where the load is inclined at an angle α to the grain, $\sigma_{c,adm,\alpha}$	$= \sigma_{c,adm,\parallel} - (\sigma_{c,adm,\parallel} - \sigma_{c,adm,\perp}) \sin \alpha$

2.6 Symbols

The following symbols and subscripts are used to identify section properties of timber elements, applied loading conditions, type of force and induced and permissible stresses. Symbols and subscripts are kept as similar as possible to those given in Part 2 of BS 5268 : 1996.

Geometrical and mechanical properties

a	distance
α	angle of grain
A	area
b	breadth of beam, thickness of member
d	diameter
E	modulus of elasticity
E_{mean}	mean value of modulus of elasticity
E_{min}	minimum value of modulus of elasticity
G	modulus of rigidity or shear modulus
h	depth of member
i	radius of gyration
I	second moment of area
L	length, span
L_e	effective length, effective span
m	mass
n	number
λ	slenderness ratio
Q	first moment of area
ρ_k	characteristic density
ρ_{mean}	average density
Z	section modulus

Bending

M	bending moment
$\sigma_{m,a,\parallel}$	applied bending stress parallel to grain
$\sigma_{m,g,\parallel}$	grade bending stress parallel to grain
$\sigma_{m,adm,\parallel}$	permissible bending stress parallel to grain

Shear

F_v	applied shear force
$\tau_{a,\parallel}$	applied shear stress parallel to grain
$\tau_{g,\parallel}$	grade shear stress parallel to grain
$\tau_{adm,\parallel}$	permissible shear stress parallel to grain
$\tau_{r,a}$	applied rolling shear stress
$\tau_{r,adm}$	permissible rolling shear stress

Deflection

Δ_m	bending deflection
Δ_s	shear deflection
Δ_{total}	total deflection due to bending and shear
Δ_{adm}	permissible deflection

Compression

$\sigma_{c,a,\parallel}$	applied compressive stress parallel to grain
$\sigma_{c,g,\parallel}$	grade compressive stress parallel to grain
$\sigma_{c,adm,\parallel}$	permissible compressive stress parallel to grain
$\sigma_{c,a,\perp}$	applied compressive stress perpendicular to grain
$\sigma_{c,g,\perp}$	grade compressive stress perpendicular to grain
$\sigma_{c,adm,\perp}$	permissible compressive stress perpendicular to grain

Tension

$\sigma_{t,a,\parallel}$	applied tensile stress parallel to grain
$\sigma_{t,g,\parallel}$	grade tensile stress parallel to grain
$\sigma_{t,adm,\parallel}$	permissible tensile stress parallel to grain

2.7   References

1. Arya, C. (1994) *Design of structural elements*. E. & F. N. Spon, London.
2. British Standards Institution (1984, 1995, 1988) BS 6399 : *Loading for buildings*. Part 1 : 1984 : *Code of practice for dead and imposed loads*. Part 2 : 1995 : *Code of practice for wind loads*. Part 3 : 1988 : *Code of practice for imposed roof loads*. BSI, London.
3. British Standards Institution (1972) CP 3 : Chapter V : *Loading*. Part 2 : 1972 : *Wind loads*. BSI, London.
4. British Standards Institution (1995) BS EN 336 : *Structural timber. Coniferous and poplar. Sizes. Permissible deviations*. BSI, London.

Chapter 3
Using Mathcad® for Design Calculations

3.1 Introduction

Many academic institutions and design offices are turning to computer assisted instructions. This is especially true in science and engineering where courses are being introduced to teach the use of computers as analysis and design tools. Mathcad's potential as a powerful and easy to use computational tool has already been recognised by most academic institutions and many design offices.

The aim of this chapter is to demonstrate how the analysis and design calculations for structural timber can be incorporated into simple to use electronic notepads or worksheets. Access to a personal computer (PC) and the associated software Mathcad is not a prerequisite for understanding the design calculations in the examples provided in this book. All design examples given are fully self-explanatory and well annotated. They have been produced in the form of worksheets to run under Mathcad, version 6, or higher, in either one of its editions, i.e. Student, Standard, Plus or Professional. Details are given at the end of this book.

The design worksheets given are intended as a source of study, practice and further development by the reader. They should not be seen as complete and comprehensive design worksheets but rather as the foundations of a design system that can be developed further. The aim is to encourage readers to use computing as a tool to increase their understanding of how design solutions vary in response to a change in one of the variables and how alternative design options can be obtained easily and effortlessly, allowing the design engineer to arrive at the most suitable/economic solution very quickly.

It is important to note that this chapter is not intended to teach Mathcad. It aims only to familiarise the reader with the Mathcad worksheet formats that are used to produce design examples in this book.

3.2 What is Mathcad?[1]

Mathcad (developed by MathSoft, Inc.) is an electronic notepad (live worksheet) that allows mathematical calculation to be performed on a computer screen in a format similar to the way it would be done manually with paper and pencil. While Mathcad employs the usual mathematical symbols (i.e. $+$, $-$, $/$, $=$) for algebraic operations, it also uses the conventional symbols of calculus for differentiation and integration to perform these operations. It preserves the conventional symbolic form for subscribing, special mathematical and trigonometrical functions, series operations, and matrix algebra. When expository text is added, Mathcad's symbolic format leads to reports that are understood easily by others. Data can be presented in both tabular and graphical forms.

Mathcad can also be used to answer, amongst many others, the 'what-if' questions in engineering problems. With a well structured worksheet, design calculations can be performed whereby parameters can be changed and the results viewed almost immediately on the computer display and/or printed.

3.3   What does Mathcad do?[2]

Mathcad combines the live document interface of a spreadsheet with the WYSIWYG interface of a word processor. With Mathcad, equations can be typeset on the screen in exactly the way they are presented in textbooks, with the advantage that it can also do the calculations.

Mathcad also comes with multiple fonts and the ability to print what you see on the screen on any Windows supported printer. This, combined with Mathcad's live document interface, makes it easy to produce up-to-date, publication-quality engineering reports and/or design solution sheets.

The following subsections demonstrate how some simple operations are carried out in Mathcad. This is to illustrate the format/meaning of the operations used to produce the examples in this text.

3.3.1   *A simple calculation[2]*

Although Mathcad can perform sophisticated mathematics, it can just as easily be used as a simple calculator. For example,

Click anywhere in the worksheet; you will see a small crosshair.

Type $15 - 8/104.5 =$

As soon as the equal sign is pressed, Mathcad computes and shows the result (see Fig. 3.1).

3.3.2   *Definitions and variables[2]*

Mathcad's power and versatility quickly becomes apparent when the variables and functions are being used. By defining variables and functions, equations can be linked together and intermediate results can be used in further calculations.

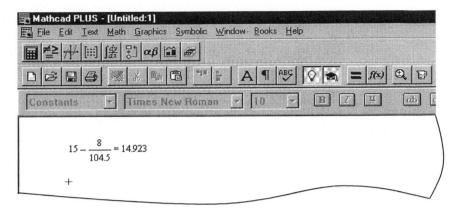

Fig. 3.1 A simple calculation.

For example, to define a value of say 10 to a variable, say *t*, click anywhere in the worksheet and type t: (the letter t followed by a colon). Mathcad will show the colon as the definition symbol := and will create an empty place holder to its right. Then type 10 in the empty placeholder to complete the definition for *t*.

To enter another definition, press [↵] to move the crosshair below the first equation. For example, to define *acc* as −9.8, type acc:−9.8. Then press [↵] again.

Now that the variables *acc* and *t* are defined, they can be used in other expressions. For example, to calculate the magnitude of $\dfrac{acc}{2} t^2$ type acc/ 2*t^2. The caret symbol ^ represents raising to a power, the asterisk * is multiplication, and the slash / is division.

To obtain the result, type =. Mathcad will return the result (as shown in Fig. 3.2).

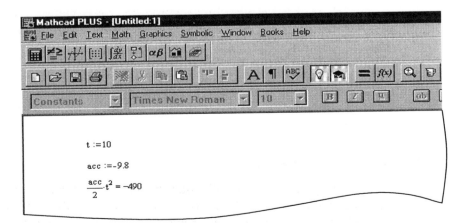

Fig. 3.2 Calculating with variables and functions.

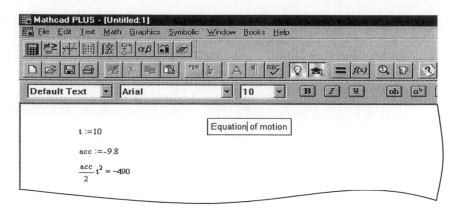

Fig. 3.3   Entering text.

3.3.3   *Entering text*[2]

Mathcad handles text as easily as it does equations. To begin typing text, click in an empty space and choose **Create Text Region** from the **Text** menu or simply click on the icon ⒜ on the menu bar. Mathcad will then create a text box in which you can type, change font, format and so on as you would when using a simple Windows based word processor. The text box will grow as the text is entered.

Now type, say, 'Equation of motion' (see Fig. 3.3). To exit text mode simply click outside the text box.

3.3.4   *Working with units*[2]

Units of measurement, while not required in Mathcad equations, can help detect and enhance the display of computed results. Mathcad's unit capabilities take care of many of the usual chores associated with using units

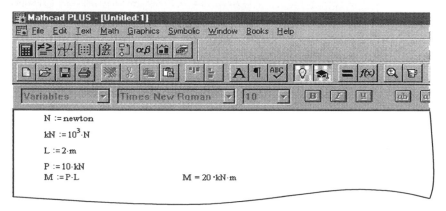

Fig. 3.4   Equations using units.

and dimensions in engineering analysis and design calculations. Once the appropriate definitions are entered, Mathcad automatically performs unit conversions and flags up incorrect and inconsistent dimensional calculations.

Although Mathcad's latest edition recognises most common units, you may wish to define your own units. For example, N = newton, and kN = 10^3 N. To assign units to a number, simply multiply the number by the name or letter(s) which defines the unit.

To illustrate this, calculate the magnitude of the bending moment M at the built-in end of a cantilever of length $L = 2$ m induced by a force of $P = 10$ kN acting at its free end. To do this, click anywhere in a Mathcad worksheet and type:

N := newton
kN := 10^3*N
L := 2*m
P := 10*kN
M := P*L

Then type M=. As soon as the = sign is typed, Mathcad will compute the result and also display the units of M (as shown in Fig. 3.4).

3.4 Summary

The previous examples aimed to demonstrate the simplicity of using Mathcad in producing the design examples given in the proceeding chapters of this book. To learn more about Mathcad, refer to the next section in this chapter.

3.5 References

1. Wieder, S. (1992) *Introduction to Mathcad for Scientists and Engineers.* McGraw-Hill, Inc., Hightstown.
2. *Mathcad User's Guide, Mathcad 6.0* (1996) MathSoft, Inc., MA.

Chapter 4
Design of Flexural Members (Beams)

4.1 Introduction

Flexural members are those subjected to bending. There are several types and forms of flexural timber members that are used in construction. Typical examples are solid section rectangular beams, floor joists, rafters and purlins. Other examples include glulam beams (vertical and horizontal glued laminated beams), ply-webbed beams (I-beams and box-beams) and beams of simple composites (Tee and I shaped beams).

Although the design principles are essentially the same for all bending members of all materials, the material characteristics are different. Steel for example is ductile, homogeneous, and isotropic. Concrete is brittle and can be assumed homogeneous for most practical purposes. As for timber, the material properties are different in the two main directions: parallel and perpendicular to the grain. Even though the normal stresses due to bending are parallel to grain direction, support conditions may impose stresses that are perpendicular to grain direction. These stresses, in addition to the primary stresses, should be checked in the design against the permissible values, which include the effects of environmental conditions, material and geometrical characteristics.

This chapter deals in detail with the general considerations necessary for the design of flexural members and describes the design details of solid section rectangular timber beams. Design methods for glued laminated beams and ply-webbed beams are described in Chapters 6 and 7, respectively.

4.2 Design considerations

The main design considerations for flexural members are:

(1) bending stress and prevention of lateral buckling
(2) deflection
(3) shear stress
(4) bearing stress.

The cross-sectional properties of all flexural members have to satisfy elastic strength and service load requirements. In general, bending is the most critical criterion for medium-span beams, deflection for long-span beams and shear for heavily loaded short-span beams. In practice, design checks are carried out for all criteria listed above.

In Chapter 2 it was mentioned that the design of timber elements, connections and components is based on the recommendations of BS 5268 : Part 2 : 1996 which is still based on 'permissible stress' design philosophy. The permissible stress value is calculated as the product of the grade stress and the appropriate modification factors for particular service and loading conditions, and is usually compared with the applied stress in a member or part of a component in structural design calculations. In general:

permissible stress (= grade stress × K-factors) ≥ applied stress

4.3 Bending stress and prevention of lateral buckling

The design of timber beams in flexure requires the application of the elastic theory of bending as expressed by:

$$\sigma = \frac{M \cdot y}{I} \tag{4.1}$$

The term I/y is referred to as section modulus and is denoted by Z. Using the notations defined in Chapter 2, the applied bending stress about the major (x–x) axis of the beam (say) (see Fig. 4.1), is calculated from:

$$\sigma_{m,a,\parallel} = \frac{M \cdot y}{I_{xx}} = \frac{M}{Z_{xx}} \tag{4.2}$$

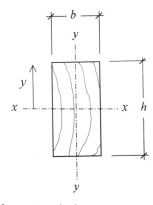

Fig. 4.1 Cross-section of a rectangular beam.

where:

$\sigma_{m,a,\parallel}$ = applied bending stress (in N/mm^2)

M = maximum bending moment (in Nmm)

Z_{xx} = section modulus about its major (x–x) axis (in mm^3). For rectangular sections

$$Z_{xx} = \frac{I_{xx}}{y} = \frac{\dfrac{bh^3}{12}}{\dfrac{h}{2}} = \frac{bh^2}{6} \tag{4.3}$$

I_{xx} = second moment of area about x–x axis (in mm^4)

y = distance from the neutral-axis of the section to the extreme fibres (in mm)

h = depth of the section (in mm)

b = width of the section (in mm).

The permissible bending stress $\sigma_{m,adm,\parallel}$ is calculated as the product of grade bending stress parallel to grain $\sigma_{m,g,\parallel}$ and any relevant modification factors (K-factors). These are K_2 for wet exposure condition (if applicable), K_3 for load-duration, K_6 for solid timber members other than rectangular sections (if applicable), K_7 for solid timber members other than 300 mm deep, and K_8 for load-sharing systems (if applicable). Hence:

$$\sigma_{m,adm,\parallel} = \sigma_{m,g,\parallel} \times K_2 \times K_3 \times K_6 \times K_7 \times K_8 \tag{4.4}$$

K_2, K_3 and K_8 are general modification factors, which were described in detail in Chapter 2. K_6 and K_7 specifically relate to the calculation of permissible bending stress, $\sigma_{m,adm,\parallel}$ and are described in the following sections.

4.3.1 *Effective span, L_e*

Clause 2.10.3 of BS 5268 : Part 2 recommends that the span of flexural members should be taken as the distance between the centres of bearings.

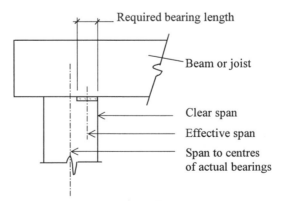

Fig. 4.2   Effective span (Baird and Ozelton[1]).

Where members extend over bearings, which are longer than is necessary, the spans may be measured between the centres of bearings of a length which should be adequate in accordance with Part 2 of the code (see Fig. 4.2).

In determining the effective span, L_e, it is usually acceptable to assume an addition of 50 mm to the clear span, between the supports, for solid timber beams and joists and 100 mm for built-up beams on spans up to around 12 m, but longer spans should be checked.[1]

4.3.2 Form factor, K_6

Grade bending stress values given in the code apply to solid timber members of rectangular cross-section. For shapes other than rectangular (see Fig. 4.3), the grade bending stress value should be multiplied by the modification factor K_6 where:

$K_6 = 1.18$ for solid circular sections, and
$K_6 = 1.41$ for solid square sections loaded diagonally (see Fig. 4.3).

4.3.3 Depth factor, K_7

The grade bending stresses given in Tables 7–12a of BS 5268 : Part 2 apply to beams having a depth, h, of 300 mm (Clause 2.10.6). For other depths of beams, the grade bending stress should be multiplied by the depth modification factor, K_7, where:

for $h \leq 72$ mm, $K_7 = 1.17$

for 72 mm $< h < 300$ mm, $K_7 = \left(\dfrac{300}{h}\right)^{0.11}$

for $h > 300$ mm, $K_7 = 0.81 \dfrac{h^2 + 92\,300}{h^2 + 56\,800}$

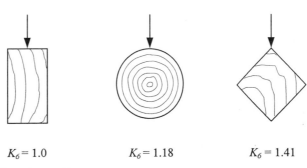

$K_6 = 1.0$ $K_6 = 1.18$ $K_6 = 1.41$

Fig. 4.3 Form factor K_6.

4.3.4 *Selection of a suitable section size*

There are two methods commonly used in selecting an appropriate trial section:

(1) Engineering judgement which is based on experience.
(2) By utilising the permissible bending stress criterion in equation (4.2). Thus the expression for calculation of the required section modulus Z_{xx} for timber members, incorporating all the relevant K-factors, is as follows:

$$Z_{xx,required} = \frac{M}{\sigma_{m,adm,/\!/}} = \frac{M}{\sigma_{m,g,/\!/} K_2 K_3 K_6 K_7 K_8} \qquad (4.5)$$

Thus a suitable section size having a $Z_{xx} \geq Z_{xx,required}$ can be selected. The chosen section should then be checked for lateral stability, deflection, shear and bearing.

The standard (customary) sizes of timber sections in the UK, normally available from stock, are given in the National Annex to BS EN 336. A summary of details is given in Appendix A.

4.3.5 *Lateral stability*

BS 5268 : Part 2 : 1996 recommends that the depth to breadth ratio of solid and laminated rectangular beams should not exceed the values given in

Table 4.1 Maximum depth to breadth ratio for solid and laminated members, (Table 16, BS 5268 : Part 2)

Degree of lateral support	Maximum depth to breadth ratio
No lateral support	2
Ends held in position	3
Ends held in position and member held in line as by purlins or tie rods at centres not more than 30 times breadth of the member	4
Ends held in position and compression edge held in line, as by direct connection of sheathing, deck or joists	5
Ends held in position and compression edge held in line, as by direct connection of sheathing, deck or joists, together with adequate bridging or blocking spaced at intervals not exceeding 6 times the depth	6
Ends held in position and both edges held firmly in line	7

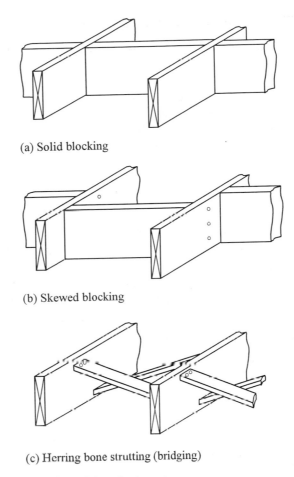

(a) Solid blocking

(b) Skewed blocking

(c) Herring bone strutting (bridging)

Fig. 4.4   Examples of provisions for lateral support.

Table 16 (reproduced here as Table 4.1) of the code corresponding to the appropriate degree of lateral support. Examples of provisions for lateral support are shown in Fig. 4.4.

4.3.6   *An illustrative example*

Determine the value of permissible bending stress parallel to grain $\sigma_{m,adm,//}$ and magnitude of maximum bending moment for a main beam of 50 mm × 200 mm deep Canadian Douglas fir-larch grade SS under service class 2 and short-duration loading.

BS 5268 : Part 2	Description	Output
Table 2	strength classification	strength class $= C24$
Table 7	grade stress // to grain	$\sigma_{m,g,/\!/} = 7.5\,\text{N/mm}^2$
Clause 1.6.4	service class 2	$K_2 = 1$
Table 14	short-duration loading	$K_3 = 1.5$
Clause 2.10.5	rectangular section	$K_6 = 1$
Clause 2.10.6	depth factor	$K_7 = \left(\dfrac{300}{200}\right)^{0.11} = 1.045$
Clause 2.9	no load-sharing	$K_8 = 1$

Permissible bending stress

$$\sigma_{m,adm,/\!/} = \sigma_{m,g,/\!/} K_2 K_3 K_6 K_7 K_8 = 7.5 \times 1.0 \times 1.5 \times 1.0 \times 1.045 \times 1.0$$

$$\sigma_{m,adm,/\!/} = 11.75\,\text{N/mm}^2$$

Allowable maximum bending moment is obtained by rearranging equation (4.5)

$$M = Z_{xx,provided} \times \sigma_{m,adm,/\!/} = \frac{bh^2}{6} \times \sigma_{m,adm,/\!/} = \frac{50 \times 200^2}{6} \times 11.75$$

$$M = 3.91\,\text{kNm}$$

4.4 Deflection

BS 5268 : Part 2, Clause 2.10.7 recommends that 'The dimensions of flexural members should be such as to restrict deflection within limits appropriate to the type of structure, having regard to the possibility of damage to surfacing materials, ceilings, partitions and to the functional needs as well as aesthetic requirements.'

4.4.1 Deflection limits

In most cases, including domestic flooring, the combined deflection due to bending, Δ_m, and shear, Δ_s, should not exceed 0.003 of the span to satisfy this recommendation. In addition, for domestic floor joists, the deflection under full load should not exceed the lesser of 0.003 times the span or 14 mm. This is to avoid undue vibration under moving or impact loading.

In general $\qquad\qquad \Delta_{total} = (\Delta_m + \Delta_s) \leq (0.003 \times \text{span})$

and for domestic joists $\quad \Delta_{total} \leq \text{lesser of } (0.003 \times \text{span or } 14\,\text{mm})$

4.4.2 *Precamber*

Subject to consideration being given to the effect of excessive deformation, timber beams may be precambered to account for the deflection under full dead or permanent load. In this instance, BS 5268 : Part 2 recommends that the deflection due to imposed load only should not exceed 0.003 of the span.

4.4.3 *Bending deflection*

The maximum bending deflection induced by the two most common load cases is given below:

(1) For a simply supported beam carrying a uniformly distributed load of W_{total}

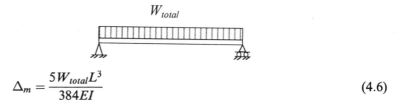

$$\Delta_m = \frac{5W_{total}L^3}{384EI} \tag{4.6}$$

(2) For a simply supported beam carrying a concentrated load at mid-span of P

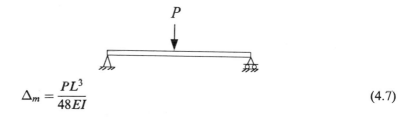

$$\Delta_m = \frac{PL^3}{48EI} \tag{4.7}$$

where:

W_{total} = total uniformly distributed load

P = concentrated load acting at mid-span

L = effective span

I = second moment of area about axis of bending, usually beam's major $(x–x)$ axis

$E = E_{min}$ for a beam acting on its own

$\quad = E_{mean}$ for a beam in a load-sharing system.

For a single-span simply supported beam subjected to a maximum bending moment of M_{max}, irrespective of the loading type, the maximum deflection of the beam may be estimated using:

$$\Delta_m \simeq \frac{0.104M_{max}L^2}{EI} \tag{4.8}$$

Table 4.2 Modification factor K_9 used to modify the minimum modulus of elasticity for trimmer joists and lintels (Table 17, BS 5268 : Part 2)

Number of pieces	Value of K_9	
	Softwoods	Hardwoods
1	1.00	1.00
2	1.14	1.06
3	1.21	1.08
4 or more	1.24	1.10

Note: BS 5268 : Part 2, Clause 2.10.11 recommends that for trimmer joists and lintels which comprise two or more pieces connected together in parallel and acting together to support the loads, the minimum modulus of elasticity modified by the modification factor K_9, given in Table 17 of the code, should be used for calculation of deflections. Table 17 is reproduced here as Table 4.2.

4.4.4 *Shear deflection*

Since in timber and wood based structural materials the shear modulus is considerably lower as a proportion of the modulus of elasticity, compared to other structural materials such as steel, the effect of shear deflection can be significant and should be considered in the design calculations.

The maximum shear deflection, Δ_s, induced in a single-span simply supported beam of either rectangular or square cross-section, may be determined from the following equation:

$$\Delta_s = \frac{19.2M_{max}}{AE} \tag{4.9}$$

where A is the cross-sectional area of the beam, M_{max} is the maximum bending moment in the beam and E is as defined above.

4.5 Bearing stress

The bearing stresses in timber beams are developed due to compressive forces applied in a direction perpendicular to the grain and occur in positions such as points of support or applied concentrated loads. Possible bearing failure positions are shown in Fig. 4.5.

The applied bearing stress, $\sigma_{c,a,\perp}$ is calculated from the following equation:

$$\sigma_{c,a,\perp} = \frac{F}{A_{bearing}} \tag{4.10}$$

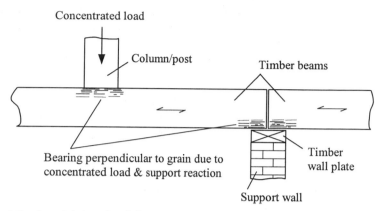

Fig. 4.5 Possible bearing failures.

where:

   F = bearing force (usually maximum reaction or concentrated load)
$A_{bearing}$ = bearing area (= bearing length × breadth of the section).

In general, the value of applied bearing stress, $\sigma_{c,a,\perp}$ should not exceed the permissible bearing stress, $\sigma_{c,adm,\perp}$ determined from:

$$\sigma_{c,adm,\perp} = \sigma_{c,g,\perp} \times K_2 \times K_3 \times K_4 \times K_8 \qquad (4.11)$$

K_2, K_3 and K_8 are general modification factors, which were described in detail in Chapter 2. K_4 relates to the calculation of permissible bearing stress, $\sigma_{c,adm,\perp}$, and is described below.

4.5.1 *Length and position of bearings*

BS 5268: Part 2, Clause 2.10.2 recommends that the grade stresses for compression perpendicular to the grain apply to bearings of any length at the ends of a member, and bearings 150 mm or more in length at any position. For bearing less than 150 mm long located 75 mm or more from the end of a member, as shown in Fig. 4.6, the grade stress should be multiplied by the modification factor K_4 given in Table 15 of the code (reproduced here as Table 4.3).

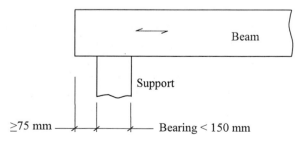

Fig. 4.6 Bearing length and position.

Table 4.3 Modification factor K_4 for bearing stress (Table 15, BS 5268: Part 2)

Length of bearing (mm)	K_4
10	1.74
15	1.67
25	1.53
40	1.33
50	1.20
75	1.14
100	1.10
150 or more	1.00

4.6   Shear stress

The critical position for shear is usually at supports where maximum reaction occurs. The applied shear stress, τ, is calculated as the maximum (not average) shear stress from the following equation:

$$\tau = \frac{F_v \cdot Q}{I \cdot b} \tag{4.12}$$

For a rectangular timber beam, the maximum applied shear stress parallel to grain, $\tau_{a,/\!/}$ occurs at the neutral axis and is calculated from:

$$\tau_{a,/\!/} = 1.5 \, \frac{F_v}{A} \tag{4.13}$$

where:

F_v = maximum vertical shear force (usually maximum reaction)
A = cross-sectional area
Q = first moment of area about neutral axis above the position where shear stress is required
I = second moment of area
b = breadth of the section at the position where shear stress is required.

In general, the value of applied shear stress, $\tau_{a,/\!/}$, should not exceed the permissible shear stress parallel to grain, $\tau_{adm,/\!/}$, determined from:

$$\tau_{adm,/\!/} = \tau_{g,/\!/} \times K_2 \times K_3 \times K_5 \times K_8 \tag{4.14}$$

K_2, K_3 and K_8 are general modification factors, which were described in detail in Chapter 2. K_5 relates to the calculation of the permissible shear stress, $\tau_{adm,/\!/}$, for members with notched ends, and is described below.

4.6.1   *Shear at notched ends*

K_5 is a modification factor which allows for stress concentration induced at square-cornered notches at the ends of a flexural member (Clause 2.10.4, BS 5268: Part 2) where:

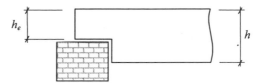

(a) Beam with notch on the underside

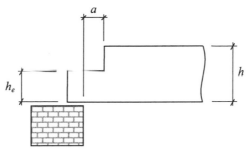

(b) Beam with notch on the top edge

Fig. 4.7   Notched beams (adapted from BS 5268 : Part 2).

(1)   For a notch on the underside of a beam, see Fig. 4.7(a),

$$K_5 = \frac{h_e}{h}$$

(2)   For a notch on the top edge, see Fig. 4.7(b),

$$K_5 = \frac{h(h_e - a) + ah_e}{h_e^2} \quad \text{for } a \le h_c$$

$$= 1.0 \qquad\qquad \text{for } a > h_e$$

4.7   Suspended timber flooring

A suspended flooring system generally comprises a series of joists closely spaced, being either simply supported at their ends or continuous over load-bearing partition walls. The floor boarding or decking is applied on the top of the joists and underneath ceiling linings are fixed. A typical suspended floor arrangement is shown in Fig. 4.8(a)

The distance between the centres of the joists is normally governed by the size of the decking and ceiling boards, which are normally available in dimensions of 1200 mm wide × 2400 mm long. The size of the decking and ceiling boards allows convenient joist spacings of 300 mm, 400 mm or 600 mm centre to centre. In addition, the choice of joist spacing may also be affected by the spanning capacity of the flooring material, joist span and other geometrical constraints such as an opening for a stairwell.

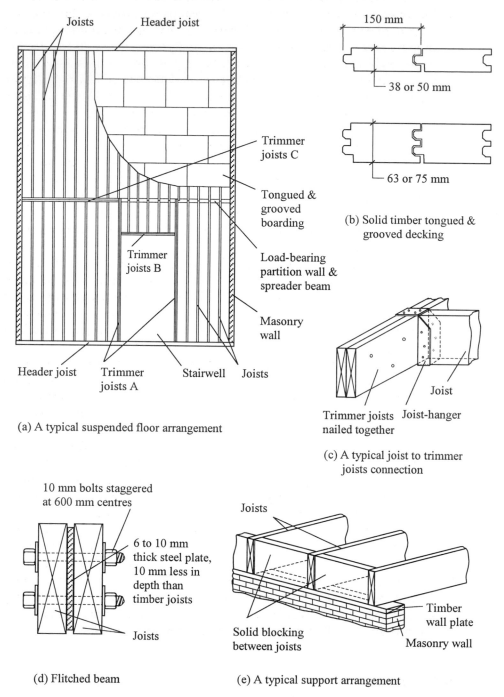

(a) A typical suspended floor arrangement

(b) Solid timber tongued & grooved decking

(c) A typical joist to trimmer joists connection

(d) Flitched beam

(e) A typical support arrangement

Fig. 4.8 Suspended timber flooring – typical components.

The most common floor decking in domestic dwellings and timber-framed buildings uses some form of wood-based panel products, for example chipboard or plywood. Solid timber decking such as softwood tongued and grooved (t & g) decking is often used in roof constructions, in conjunction with glued-laminated members, to produce a pleasant, natural timber ceiling with clear spans between the main structural members. The solid timber tongued and grooved boards are normally machined from 150 mm wide sections with 38–75 mm basic thicknesses [Fig. 4.8(b)].

The supports for joists are provided in various forms depending on the type of construction. Timber wall plates are normally used to support joists on top of masonry walls and foundations, Fig. 4.8(e). In situations where joists are to be supported on load-bearing timber-frame walls or internal partitions, header beams or spreader members are provided to evenly distribute the vertical loads. Joist-hangers are often used to attach and support joists onto the main timber beams, trimmer members or masonry walls [Fig. 4.8(c)].

Timber trimmer joists are frequently used within timber floors of all types of domestic buildings, see Fig. 4.8(a). There are two main reasons for which trimmer joists may be provided[2]. First is to trim around an opening such as a stairwell or loft access (Trimmer joists A), and to support incoming joists (Trimmer joists B), and second is to reduce the span of floor joists over long open spans (Trimmer joists C), as shown in Fig. 4.8(a).

Trimming around openings can usually be achieved by using two or more joists nailed together to form a trimmer beam, Fig. 4.8(c), or by using a single but larger timber section if construction geometry permits. Alternatively, trimmers can be of hardwood or glued laminated timber, boxed ply-webbed beams, or composite timber and steel flitched beams[2], Fig. 4.8(d).

All flooring systems are required to have fire resistance from the floor below and this is achieved by the ceiling linings, the joists and the floor boarding acting together as a composite construction[3]. For example, floors in two storey domestic buildings require modified 30 minutes fire resistance (30 minutes load-bearing, 15 minutes integrity and 15 minutes insulation). In general a conventional suspended timber flooring system comprising 12.5 mm plasterboard taped and filled, tongued and grooved floor boarding with at least 16 mm thickness directly nailed to floor joists, meets the requirements for the modified 30 minutes fire resistance provided that where joist-hangers are used they are formed from at least 1 mm thick steel of strap or shoe type. Further details and specific requirements for fire resistance are given in BS 5268 : Part 4: 'Fire resistance of timber structures'.

4.8   References

1.   Baird and Ozelton (1984) *Timber Designer's Manual*, 2nd edn. BSP Professional Books, Oxford.

2. The Swedish Finnish Timber Council (1988) *Principles of Timber Framed Construction*, Retford.
3. TRADA (1994) *Timber Frame Construction*, 2nd edn. Timber Research and Development Association (TRADA), High Wycombe.

4.9   Design examples

Example 4.1   Design of a main beam

A main beam of 3 m length spans over an opening 2.8 m wide (Fig. 4.9) and supports a flooring system which exerts a long-duration loading of 3.9 kN/m, including its own self-weight, over its span. The beam is supported by 50 mm wide walls on either side. Carry out design checks to show that a 75 mm × 225 mm deep sawn section whitewood grade SS under service class 1 is suitable.

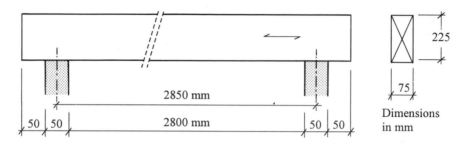

Fig. 4.9 Beam details (Example 4.1).

Definitions

Force, kN	$N :=$ newton
Length, m	$kN := 10^3 \cdot N$
Cross-sectional dimensions, mm	Direction parallel to grain, //
Stress, Nmm^{-2}	Direction perpendicular to grain, *pp*

1.   *Geometrical properties*

Span (clear distance), *L* $L := 2.8 \cdot m$

Bearing width, *bw* $bw := 50 \cdot mm$

Effective span, L_e $L_e := L + bw$

$L_e = 2.85 \circ m$

Beam dimensions:

Breadth of the section, *b* $b := 75 \cdot mm$

Depth of the section, *h* $h := 225 \cdot mm$

Cross-sectional area, A

$A := b \cdot h$
$A = 16\,875 \circ \text{mm}^2$

Second moment of area, I_{xx}

$I_{xx} := \frac{1}{12} \cdot b \cdot h^3$
$I_{xx} = 7.12 \times 10^7 \circ \text{mm}^4$

2. Loading

Applied uniformly distributed load, w
Total load, W

$w := 3.9 \cdot \text{kN} \cdot \text{m}^{-1}$
$W := w \cdot L_e$
$W = 11.11 \circ \text{kN}$

3. K-factors

Service class 1 (K_2, Table 13)
Load duration (K_3, Table 14)
Bearing: 50 mm, but located < 75 mm
   from end of member (K_4, Table 15)
Notched end effect (K_5, Clause 2.10.4)
Form factor (K_6, Clause 2.10.5)

Depth factor (K_7, Clause 2.10.6)

No load sharing (K_8, Clause 2.9)

$K_2 := 1$
$K_3 := 1$ for long-term
$K_4 := 1$

$K_5 := 1$ for no notch
$K_6 := 1$
$K_7 := \left(\frac{300 \cdot \text{mm}}{h} \right)^{0.11}$
$K_7 = 1.03$
$K_8 := 1$

4. Grade stresses

BS 5268 : Part 2, Tables 2 and 7
Whitewood grade SS
Bending parallel to grain
Compression perpendicular to grain
Shear parallel to grain
Minimum modulus of elasticity

Strength class = C24
$\sigma_{m \cdot g \cdot //} := 7.5 \cdot \text{N} \cdot \text{mm}^{-2}$
$\sigma_{c \cdot g \cdot pp} := 2.4 \cdot \text{N} \cdot \text{mm}^{-2}$ no wane
$\tau_{g \cdot //} := 0.71 \cdot \text{N} \cdot \text{mm}^{-2}$
$E_{min} := 7200 \cdot \text{N} \cdot \text{mm}^{-2}$

5. Bending stress

Applied bending moment

$M := \frac{W \cdot L_e}{8}$
$M = 3.96 \circ \text{kN} \cdot \text{m}$

Section modulus

$Z_{provided} := \frac{b \cdot h^2}{6}$
$Z_{provided} = 632\,812.5 \circ \text{mm}^3$

Applied bending stress

$\sigma_{m.a.//} := \frac{M}{Z_{provided}}$
$\sigma_{m.a.//} = 6.26 \circ \text{N} \cdot \text{mm}^{-2}$

Permissible bending stress

$$\sigma_{m.adm.//} := \sigma_{m.g.//} \cdot K_2 \cdot K_3 \cdot K_6 \cdot K_7 \cdot K_8$$
$$\sigma_{m.adm.//} = 7.74 \circ N \cdot mm^{-2}$$
Bending stress satisfactory

6. Lateral stability

BS 5268 : Part 2, Clause 2.10.8 and Table 16

Maximum depth to breadth ratio, h/b

$$\frac{h}{b} = 3$$

Ends should be held in position

7. Shear stress

Applied shear force

$$F_v := \frac{W}{2}$$
$$F_v = 5.56 \circ kN$$

Applied shear stress

$$\tau_a := \frac{3}{2} \cdot \left(\frac{F_v}{b \cdot h} \right)$$
$$\tau_a = 0.49 \circ N \cdot mm^{-2}$$

Permissible shear stress, no notch

$$\tau_{adm.//} := \tau_{g.//} \cdot K_2 \cdot K_3 \cdot K_5 \cdot K_8$$
$$\tau_{adm.//} = 0.71 \circ N \cdot mm^{-2}$$
Shear stress satisfactory

8. Bearing stress

Applied load

$$F_v := \frac{W}{2}$$
$$F_v = 5.56 \circ kN$$

Applied bearing stress

$$\sigma_{c.a.pp} := \left(\frac{F_v}{b \cdot bw} \right)$$
$$\sigma_{c.a.pp} = 1.48 \circ N \cdot mm^{-2}$$

Permissible shearing stress

$$\sigma_{c.adm.pp} := \sigma_{c.g.pp} \cdot K_2 \cdot K_3 \cdot K_4 \cdot K_8$$
$$\sigma_{c.adm.pp} = 2.4 \circ N \cdot mm^{-2}$$
Bearing stress satisfactory

9. Deflection

No load sharing

$$E := E_{min}$$

Deflection due to bending

$$\Delta_m := \frac{5 \cdot W \cdot L_e^3}{384 \cdot E \cdot I_{xx}}$$
$$\Delta_m = 6.54 \circ mm$$

Deflection due to shear

$$\Delta_s := \frac{19.2 \cdot M}{(b \cdot h) \cdot E}$$
$$\Delta_s = 0.63 \circ mm$$

Total deflection

$$\Delta_{total} := \Delta_m + \Delta_s$$
$$\Delta_{total} = 7.16 \circ \text{mm}$$

Permissible deflection

$$\Delta_{adm} := 0.003 \cdot L_e$$
$$\Delta_{adm} = 8.55 \circ \text{mm}$$
Deflection satisfactory

Therefore a 75 mm × 225 mm sawn section whitewood C24 is satisfactory

Example 4.2   Design of floor joists

A timber floor spanning 3.8 m centre to centre is to be designed using timber joists at 400 mm centres. The floor is subjected to a domestic imposed load of 1.5 kN/m² and carries a dead loading, including self-weight of 0.35 kN/m². Carry out design checks to show that a series of 44 mm × 200 mm deep sawn section British spruce grade SS under service class 1 is suitable.

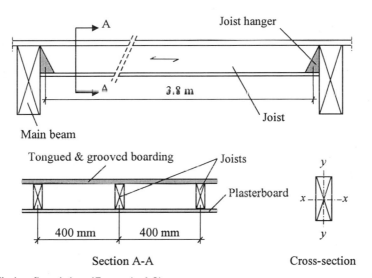

Fig. 4.10   Timber floor joists (Example 4.2).

Definitions

Force, kN

Length, m

Cross-sectional dimensions, mm

Stress, Nmm⁻²

$\text{N} := \text{newton}$

$\text{kN} := 10^3 \cdot \text{N}$

Direction parallel to grain, //

Direction perpendicular to grain, *pp*

1.   Geometrical properties

Effective span, L_e

Joist spacing, *Js*

$L_e := 3.8 \cdot \text{m}$

$Js := 0.4 \cdot \text{m}$

Joist dimensions:

Breadth of section, b

$b := 44 \cdot mm$

Depth of section, h

$h := 200 \cdot mm$

Cross-sectional area, A

$A := b \cdot h$

$A = 8.8 \times 10^3 \circ mm^2$

Second moment of area, I_{xx}

$I_{xx} := \frac{1}{12} \cdot b \cdot h^3$

$I_{xx} = 2.93 \times 10^7 \circ mm^4$

2. Loading

Dead load, DL

$DL := 0.35 \cdot kN \cdot m^{-2}$

Imposed load, IL

$IL := 1.5 \cdot kN \cdot m^{-2}$

Total load, W

$W := (DL + IL) \cdot Js \cdot L_e$

$W = 2.81 \circ kN$

3. K-factors

Service class 1 (K_2, Table 13)

$K_2 := 1$

Load duration (K_3, Table 14)

$K_3 := 1$ for long-term

Bearing: assume 50 mm, but located $< 75\,mm$ from the end of the member (K_4, Table 15)

$K_4 := 1$

Notched end effect (K_5, Clause 2.10.4)

$K_5 := 1$ for no notch

Form factor (K_6, Clause 2.10.5)

$K_6 := 1$

Depth factor (K_7, Clause 2.10.6)

$K_7 := \left(\frac{300 \cdot mm}{h} \right)^{0.11}$

$K_7 = 1.05$

Load sharing applies (K_8, Clause 2.9)

$K_8 := 1.1$

4. Grade stresses

BS 5268 : Part 2, Tables 2 and 7
British spruce grade SS

Strength class $= C18$

Bending parallel to grain

$\sigma_{m.g.//} := 5.8 \cdot N \cdot mm^{-2}$

Compression perpendicular to grain

$\sigma_{c.g.pp} := 2.2 \cdot N \cdot mm^{-2}$ no wane

Shear parallel to grain

$\tau_{g.//} := 0.67 \cdot N \cdot mm^{-2}$

Mean modulus of elasticity, load sharing

$E_{mean} := 9100 \cdot N \cdot mm^{-2}$

5. Bending stress

Applied bending moment

$M := \frac{W \cdot L_e}{8}$

$M = 1.34 \circ kN \cdot m$

Section modulus

$Z_{provided} := \frac{b \cdot h^2}{6}$

$Z_{provided} = 2.93 \times 10^5 \circ mm^3$

Applied bending stress

$$\sigma_{m.a.\|} := \frac{M}{Z_{provided}}$$

$$\sigma_{m.a.\|} = 4.55 \circ N \cdot mm^{-2}$$

Permissible bending stress

$$\sigma_{m.adm.\|} := \sigma_{m.g.\|} \cdot K_2 \cdot K_3 \cdot K_6 \cdot K_7 \cdot K_8$$

$$\sigma_{m.adm.\|} = 6.67 \circ N \cdot mm^{-2}$$

Bending stress satisfactory

6. *Lateral stability*

BS 5268 : Part 2, Clause 2.10.8, and Table 16

Maximum depth to breadth ratio, h/b

$$\frac{h}{b} = 4.55$$

Ends should be held in position and compression edges held in line

7. *Shear stress*

Applied shear force

$$F_v := \frac{W}{2}$$

$$F_v = 1.41 \circ kN$$

Applied shear stress

$$\tau_a := \frac{3}{2} \cdot \left(\frac{F_v}{b \cdot h} \right)$$

$$\tau_a = 0.24 \circ N \cdot mm^{-2}$$

Permissible shear stress, no notch

$$\tau_{adm.\|} := \tau_{g.\|} \cdot K_2 \cdot K_3 \cdot K_5 \cdot K_8$$

$$\tau_{adm.\|} = 0.74 \circ N \cdot mm^{-2}$$

Shear stress satisfactory

8. *Bearing stress*

Applied load

$$F_v := \frac{W}{2}$$

$$F_v = 1.41 \circ kN$$

Assume bearing widths, *bw*, of 50 mm either side

$$bw := 50 \cdot mm$$

Applied bearing stress

$$\sigma_{c.a.pp} := \left(\frac{F_v}{b \cdot bw} \right)$$

$$\sigma_{c.a.pp} = 0.64 \circ N \cdot mm^{-2}$$

Permissible bearing stress

$$\sigma_{c.adm.pp} := \sigma_{c.g.pp} \cdot K_2 \cdot K_3 \cdot K_4 \cdot K_8$$

$$\sigma_{c.adm.pp} = 2.42 \circ N \cdot mm^{-2}$$

Bearing stress satisfactory

9. Deflection

Load sharing system

$$E := E_{mean}$$

Deflection due to bending

$$\Delta_m := \frac{5 \cdot W \cdot L_e^3}{384 \cdot E \cdot I_{xx}}$$

$$\Delta_m = 7.53 \circ mm$$

Deflection due to shear

$$\Delta_s := \frac{19.2 \cdot M}{(b \cdot h) \cdot E}$$

$$\Delta_s = 0.32 \circ mm$$

Total deflection

$$\Delta_{total} := \Delta_m + \Delta_s$$

$$\Delta_{total} = 7.85 \circ mm$$

Permissible deflection

$$\Delta_{adm} := 0.003 \cdot L_e$$

$$\Delta_{adm} = 11.4 \circ mm$$

Deflection satisfactory

Therefore 44 mm × 200 mm sawn sections in C18 timber are satisfactory

Example 4.3   Design of floor joists – selection of a suitable section and design for notched ends

The cross-section of a suspended timber flooring system is shown in Fig. 4.11. It consists of tongued and grooved (t & g) boarding with a self-weight of $0.15\,kN/m^2$ and carries a plasterboard ceiling of $0.2\,kN/m^2$. The floor has an effective span of 4.0 m and is subjected to a domestic imposed load of $1.5\,kN/m^2$. Design the timber floor joists using timber in strength class C18 under service class 1.

If the joists are to be notched at bearings with a 72 mm deep notch, check that the notched section is also adequate.

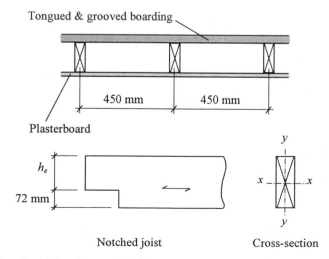

Tongued & grooved boarding

450 mm   450 mm

Plasterboard

h_e

72 mm

Notched joist

y

x — — x

y

Cross-section

Fig. 4.11   Timber floor joists (Example 4.3).

Definitions

Force, kN	$N := \text{newton}$
Length, m	$kN := 10^3 \cdot N$
Cross-sectional dimensions, mm	Direction parallel to grain, $/\!/$
Stress, Nmm^{-2}	Direction perpendicular to grain, pp

1. Geometrical properties

Effective span, L_e	$L_e := 4.0 \cdot m$
Joist spacing, Js	$Js := 0.45 \cdot m$

2. Loading

Dead load:

t & g boarding (kN/m^2), tg	$tg := 0.15 \cdot kN \cdot m^{-2}$
Plasterboard ceiling (kN/m^2), Pb	$Pb := 0.2 \cdot kN \cdot m^{-2}$
Self-weight (kN/m^2), Swt	$Swt := 0.10 \cdot kN \cdot m^{-2}$ assumed
Imposed load (kN/m^2), IL	$IL := 1.5 \cdot kN \cdot m^{-2}$
Total load (kN), W	$W := (tg + Pb + Swt + IL) \cdot Js \cdot L_e$
	$W = 3.51 \circ kN$

3. K-factors

Service class 1 (K_2, Table 13)	$K_2 := 1$
Load duration (K_3, Table 14)	$K_3 := 1$ for long-term
Bearing (K_4, Table 15), assume	$K_4 := 1$
Notched end effect (K_5, Clause 2.10.4) for no notch	$K_5 := 1$
Form factor (K_6, Clause 2.10.5)	$K_6 := 1$
Depth factor (K_7, Clause 2.10.6)	At this stage ignore K_7
Load sharing applies (K_8, Clause 2.9)	$K_8 := 1.1$

4. Grade stresses

BS 5268 : Part 2, Table 7
Strength class $= C18$

Bending parallel to grain	$\sigma_{m.g.//} := 5.8 \cdot N \cdot mm^{-2}$
Compression perpendicular to grain	$\sigma_{c.g.pp} := 2.2 \cdot N \cdot mm^{-2}$ no wane
Shear parallel to grain	$\tau_{g.//} := 0.67 \cdot N \cdot mm^{-2}$
Mean modulus of elasticity, load sharing	$E_{mean} := 9100 \cdot N \cdot mm^{-2}$

5. *Bending stress*

Applied bending moment

$$M := \frac{W \cdot L_e}{8}$$

$$M = 1.75 \circ \text{kN} \cdot \text{m}$$

Permissible bending stress

$$\sigma_{m.adm.\parallel} := \sigma_{m.g.\parallel} \cdot K_2 \cdot K_3 \cdot K_6 \cdot K_8$$

$$\sigma_{m.adm.\parallel} = 6.38 \circ \text{N} \cdot \text{mm}^{-2}$$

Required section modulus

$$Z_{required} := \frac{M}{\sigma_{m.adm.\parallel}}$$

$$Z_{required} = 2.75 \circ 10^5 \cdot \text{mm}^3$$

BS 5268 : Part 2, Clause 2.10.8 and Table 16
In order to achieve lateral stability by direct fixing of decking to joists, the depth to breadth ratio should be limited to 5, i.e. $h \le 5b$. Substituting for $h = 5b$ in $Z_{xx} = bh^2/6$ and equating it to $Z_{required}$, gives:

$$Z_{required} = \frac{b \cdot (5 \cdot b)^2}{6}$$

Thus, minimum breadth of section

$$b := \left(\frac{6 \cdot Z_{required}}{5^2} \right)^{1/3}$$

$$b = 40.42 \circ \text{mm}$$

and depth of section

$$h := 5 \cdot b$$

$$h = 202.08 \circ \text{mm}$$

Selecting a trial section from Table A2, Appendix A. Try 47 mm × 200 mm deep section.

Beam dimensions
Depth , h $h := 200 \cdot \text{mm}$
Breadth, b $b := 47 \cdot \text{mm}$

Section modulus

$$Z_{provided} := \frac{b \cdot h^2}{6}$$

$$Z_{provided} = 3.13 \times 10^5 \circ \text{mm}^3$$

Modification factor K_7 for section depth

$$K_7 := \left(\frac{300 \cdot \text{mm}}{h} \right)^{0.11}$$

$$K_7 = 1.05$$

Actual permissible bending stress

$$\sigma_{m.adm.\parallel} := \sigma_{m.g.\parallel} \cdot K_2 \cdot K_3 \cdot K_6 \cdot K_7 \cdot K_8$$

$$\sigma_{m.adm.\parallel} = 6.67 \circ \text{N} \cdot \text{mm}^{-2}$$

Bending stress satisfactory

Check self-weight
BS 5268 : Part 2, Table 7
Average density
Total joist self-weight

$$\rho := 380 \cdot \text{kg} \cdot \text{m}^{-3}$$

$$Swt_{actual} := \rho \cdot g \cdot h \cdot b \cdot L_e$$

$$Swt_{actual} = 0.14 \circ \text{kN}$$

Assumed total self-weight

$$Swt_{assumed} := Swt \cdot L_e \cdot Js$$

$$Swt_{assumed} = 0.18 \circ \text{kN}$$

Self-weight satisfactory

6. *Shear stress*

Applied shear force

$$F_v := \frac{W}{2}$$

$$F_v = 1.75 \circ kN$$

Applied shear stress

$$\tau_a := \frac{3}{2} \cdot \left(\frac{F_v}{b \cdot h} \right)$$

$$\tau_a = 0.28 \circ N \cdot mm^{-2}$$

Permissible shear stress

(1) No notch

$$\tau_{adm.\parallel} := \tau_{g.\parallel} \cdot K_2 \cdot K_3 \cdot K_5 \cdot K_8$$
$$\tau_{adm.\parallel} = 0.74 \circ N \cdot mm^{-2}$$

(2) With notch 72 mm deep

$$notch := 72 \cdot mm$$

Effective section depth

$$h_e := h - notch$$
$$h_e = 128 \circ mm$$

BS 5268 : Part 2, Clause 2.10.4

$$K_5 := \frac{h_e}{h}$$

$$\tau_{adm.\parallel} := \tau_{g.\parallel} \cdot K_2 \cdot K_3 \cdot K_5 \cdot K_8$$
$$\tau_{adm.\parallel} = 0.47 \circ N \cdot mm^{-2}$$
Shear stress satisfactory

7. *Bearing stress*

Applied load

$$F_v := \frac{W}{2}$$

$$F_v = 1.75 \circ kN$$

Permissible bearing stress

$$\sigma_{c.adm.pp} := \sigma_{c.g.pp} \cdot K_2 \cdot K_3 \cdot K_4 \cdot K_8$$
$$\sigma_{c.adm.pp} = 2.42 \circ N \cdot mm^{-2}$$

Minimum required bearing width

$$bw_{required} := \frac{F_v}{b \cdot \sigma_{c.adm.pp}}$$

$$bw_{required} = 15.43 \circ mm$$

8. *Deflection*

Second moment of area, I_{xx}

$$I_{xx} := \tfrac{1}{12} \cdot b \cdot h^3$$
$$I_{xx} = 3.13 \times 10^7 \circ mm^4$$

Load sharing system

$$E := E_{mean}$$

Deflection due to bending

$$\Delta_m := \frac{5 \cdot W \cdot L_e^3}{384 \cdot E \cdot I_{xx}}$$

$$\Delta_m = 10.26 \circ mm$$

Deflection due to shear

$$\Delta_s := \frac{19.2 \cdot M}{(b \cdot h) \cdot E}$$

$$\Delta_s = 0.39 \circ mm$$

Total deflection

$$\Delta_{total} := \Delta_m + \Delta_s$$
$$\Delta_{total} = 10.65 \circ mm$$

Permissible deflection

$$\Delta_{adm} := 0.003 \cdot L_e$$
$$\Delta_{adm} = 12 \circ mm$$

Deflection satisfactory

Therefore 47 mm × 200 mm sawn sections in C18 timber are satisfactory

Example 4.4   Design of a flooring system – floor boards and joists

The ground floor of a shop is to comprise a series of timber joists at 600 mm centres with tongued and grooved (t & g) boarding. The joists are simply supported on 100 mm hangers attached to load-bearing walls 4.2 m apart as shown in Fig. 4.12. Determine a suitable thickness for floor boarding using timber in strength class C18 and a suitable size for joists using timber in strength class C22 under service class 2. Assume imposed load is 2.0 kN/m².

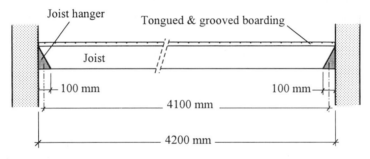

Fig. 4.12   A flooring system for Example 4.4.

Definitions

Force, kN	$N :=$ newton
Length, m	$kN := 10^3 \cdot N$
Cross-sectional dimensions, mm	Direction parallel to grain, //
Stress, Nmm^{-2}	Direction perpendicular to grain, pp

A.   Design of tongued and grooved boarding

To calculate flooring thickness boards may be designed as simply supported beams. Calculations need only be made for bending strength and deflection as span to thickness ratios and bearing widths are such that the shear and bearing stresses at supports are unlikely to be critical.

Assuming tongued and grooved boarding comprises 100 mm wide timber beams of thickness (depth) t simply supported on joists.

1. Geometrical properties

Joist spacing, Js $Js := 0.6 \cdot m$
Assume t & g boarding width, b $b := 100 \cdot mm$
Effective span, L_e $L_e := Js$

2. Loading

t & g boarding, tg $tg := 0.1 \cdot kN \cdot m^{-2}$
Imposed load, IL $IL := 2.0 \cdot kN \cdot m^{-2}$
Total load, W $W := (tg + IL) \cdot b \cdot L_e$
     $W = 0.13 \circ kN$

3. K-factors

Service class 2 (K_2, Table 13) $K_2 := 1$
Load duration (K_3, Table 14) $K_3 := 1$ for long-term
Bearing (K_4, Clause 2.10.2) $K_4 := 1$ assumed
Notched end effect (K_5, Clause 2.10.4) $K_5 := 1$ for no notch
Form factor (K_6, Clause 2.10.5) $K_6 := 1$
Depth factor (K_7, Clause 2.10.6) $K_7 := 1.17$
   for $h \leq 72\,mm$
Load sharing applies (K_8, Clause 2.9) $K_8 := 1.1$

4. Grade stresses

BS 5268 : Part 2, Tables 2 and 7
Strength class $= C18$
Bending parallel to grain $\sigma_{m.g.//} := 5.8 \cdot N \cdot mm^{-2}$
Compression perpendicular to grain $\sigma_{c.g.pp} := 2.2 \cdot N \cdot mm^{-2}$ no wane
Shear parallel to grain $\tau_{g.//} := 0.67 \cdot N \cdot mm^{-2}$
Mean modulus of elasticity, load sharing $E_{mean} := 9100 \cdot N \cdot mm^{-2}$

5. Bending stress

Applied bending moment $M := \dfrac{W \cdot L_e}{8}$

     $M = 9.45 \times 10^{-3} \circ kN \cdot m$

Permissible bending stress $\sigma_{m.adm.//} := \sigma_{m.g.//} \cdot K_2 \cdot K_3 \cdot K_6 \cdot K_7 \cdot K_8$
     $\sigma_{m.adm.//} = 7.46 \circ N \cdot mm^{-2}$

Required section modulus $Z_{required} := \dfrac{M}{\sigma_{m.adm.//}}$

     $Z_{required} = 1.27 \times 10^3 \circ mm^3$

Required thickness for t & g can be
   obtained by rearranging $Z := \dfrac{b \cdot t^2}{6}$ $t := \sqrt{\dfrac{6 \cdot Z_{required}}{b}}, \; t = 8.72 \circ mm$

6. *Deflection*

Load sharing system
$$E := E_{mean}$$

Permissible deflection
$$\Delta_{adm} := 0.003 \cdot L_e$$
$$\Delta_{adm} = 1.8 \circ mm$$

Using
$$\Delta := \frac{5 \cdot W \cdot L_e^3}{384 \cdot E \cdot I_{xx}}$$

$$I_{xx} := \frac{5 \cdot W \cdot L_e^3}{384 \cdot E \cdot \Delta_{adm}}$$

$$I_{xx} = 2.16 \times 10^4 \circ mm^4$$

Therefore from $I_{xx} = \dfrac{b \cdot t^3}{12}$
$$t := \sqrt[3]{\frac{12 \cdot I_{xx}}{b}}$$

$$t = 13.74 \circ mm$$

Therefore $t \geq$ (the greater of 13.74 mm and 8.72 mm and allowing for wear), thus
$$t := 16 \cdot mm$$

Adopt 16 mm t & g boarding using strength class C18 timber

B *Design of floor joists*

7. *Geometrical properties*

Effective span, L_e $L_e := 4.1 \cdot m$

Joist spacing, Js $Js := 0.6 \cdot m$

Bearing width, bw $bw := 100 \cdot m$

8. *Loading*

Average density (Table 7) $\rho := 410 \cdot kg \cdot m^{-3}$

t & g boarding, tg $tg := \rho \cdot g \cdot t$
$$tg = 0.06 \circ kN \cdot m^{-2}$$

Self-weight of each joist (kN/m²), Swt $Swt := 0.10 \cdot kN \cdot m^{-2}$ assumed

Imposed load (kN/m²), IL $IL := 2.0 \cdot kN \cdot m^{-2}$

Total load (kN), W $W := (tg + Swt + IL) \cdot Js \cdot L_e$
$$W = 5.32 \circ kN$$

9. *K-factors*

Service class 2 (K_2, Table 13) $K_2 := 1$

Load duration (K_3, Table 14) $K_3 := 1$ for long-term

Bearing (K_4, Table 15), assume $K_4 := 1$

Notched end effect (K_5, Clause 2.10.4) $K_5 := 1$ for no notch

Form factor (K_6, Clause 2.10.5) $K_6 := 1$

Depth factor (K_7, Clause 2.10.6)   At this stage ignore K_7

Load sharing applies (K_8, Clause 2.9) $K_8 := 1.1$

10. *Grade stresses*

BS 5268 : Part 2, Table 7
Strength class $= C22$
Bending parallel to grain $\sigma_{m.g.\//} := 6.8 \cdot N \cdot mm^{-2}$
Compression perpendicular to grain $\sigma_{c.g.pp} := 2.3 \cdot N \cdot mm^{-2}$ no wane
Shear parallel to grain $\tau_{g.\//} := 0.71 \cdot N \cdot mm^{-2}$
Mean modulus of elasticity, load sharing $E_{mean} := 9700 \cdot N \cdot mm^{-2}$

11. *Bending stress*

Applied bending moment

$$M := \frac{W \cdot L_e}{8}$$

$$M = 2.73 \circ kN \cdot m$$

Permissible bending stress

$$\sigma_{m.adm.\//} := \sigma_{m.g.\//} \cdot K_2 \cdot K_3 \cdot K_6 \cdot K_8$$
$$\sigma_{m.adm.\//} = 7.48 \circ N \cdot mm^{-2}$$

Required section modulus

$$Z_{required} := \frac{M}{\sigma_{m.adm.\//}}$$

$$Z_{required} = 3.65 \times 10^5 \circ mm^3$$

BS 5268 : Part 2, Clause 2.10.8 and Table 16
In order to achieve lateral stability by direct fixing of decking to joists, the depth to breadth ratio should be limited to 5, i.e. $h \le 5b$. Substituting for $h = 5b$ in $Z_{xx} = bh^2/6$ and equating it to $Z_{required}$, gives:

$$Z_{required} = \frac{b \cdot (5 \cdot b)^2}{6}$$

Thus, minimum breadth of section

$$b := \left(\frac{6 \cdot Z_{required}}{5^2} \right)^{1/3}$$

$$b = 44.4 \circ mm$$

and depth of section

$$h := 5 \cdot b$$
$$h = 222.02 \circ mm$$

Selecting a trial section from Table A2, Appendix A. Try 47 mm × 225 mm deep section.

Beam dimensions
Depth, h $h := 225 \cdot mm$
Breadth, b $b := 47 \cdot mm$

Section modulus

$$Z_{provided} := \frac{b \cdot h^2}{6}$$

$$Z_{provided} = 3.97 \times 10^5 \circ mm^3$$

Modification factor K_7 for section depth

$$K_7 := \left(\frac{300 \cdot mm}{h} \right)^{0.11}$$

$$K_7 = 1.03$$

Actual permissible bending stress

$$\sigma_{m.adm.//} := \sigma_{m.g.//} \cdot K_2 \cdot K_3 \cdot K_6 \cdot K_7 \cdot K_8$$
$$\sigma_{m.adm.//} = 7.72 \circ N \cdot mm^{-2}$$
Bending stress satisfactory

Check self-weight
BS 5268 : Part 2, Table 7
Average density

$$\rho := 410 \cdot kg \cdot m^{-3}$$

Total joist self-weight

$$Swt_{actual} := \rho \cdot g \cdot h \cdot b \cdot L_e$$
$$Swt_{actual} = 0.17 \circ kN$$

Assumed total self-weight

$$Swt_{assumed} := Swt \cdot L_e \cdot Js$$
$$Swt_{assumed} = 0.25 \circ kN$$
Self-weight satisfactory

12. *Shear stress*

Applied shear force

$$F_v := \frac{W}{2}$$
$$F_v = 2.66 \circ kN$$

Applied shear stress

$$\tau_a := \frac{3}{2} \cdot \left(\frac{F_v}{b \cdot h} \right)$$
$$\tau_a = 0.38 \circ N \cdot mm^{-2}$$

Permissible shear stress

$$\tau_{adm.//} := \tau_{g.//} \cdot K_2 \cdot K_3 \cdot K_5 \cdot K_8$$
$$\tau_{adm.//} = 0.78 \circ N \cdot mm^{-2}$$
Shear stress satisfactory

13. *Bearing stress*

Applied load

$$F_v := \frac{W}{2}$$
$$F_v = 2.66 \circ kN$$

Permissible bearing stress

$$\sigma_{c.adm.pp} := \sigma_{c.g.pp} \cdot K_2 \cdot K_3 \cdot K_4 \cdot K_8$$
$$\sigma_{c.adm.pp} = 2.53 \circ N \cdot mm^{-2}$$

Minimum required bearing width

$$bw_{required} := \frac{F_v}{b \cdot \sigma_{c.adm.pp}}$$
$$bw_{required} = 22.39 \circ mm$$
$< 100\,mm$ provided, OK

14. *Deflection*

Second moment of area, I_{xx}

$$I_{xx} := \tfrac{1}{12} \cdot b \cdot h^3$$
$$I_{xx} = 4.46 \times 10^7 \circ mm^4$$

Load sharing system

$$E := E_{mean}$$

Deflection due to bending

$$\Delta_m := \frac{5 \cdot W \cdot L_e^3}{384 \cdot E \cdot I_{xx}}$$
$$\Delta_m = 11.04 \circ mm$$

Deflection due to shear

$$\Delta_s := \frac{19.2 \cdot M}{(b \cdot h) \cdot E}$$

$$\Delta_s = 0.51 \circ \text{mm}$$

Total deflection

$$\Delta_{total} := \Delta_m + \Delta_s$$

$$\Delta_{total} = 11.55 \circ \text{mm}$$

Permissible deflection

$$\Delta_{adm} := 0.003 \cdot L_e$$

$$\Delta_{adm} = 12.3 \circ \text{mm}$$

Deflection satisfactory

Therefore 47 mm × 225 mm sawn sections in C22 timber are satisfactory

Chapter 5
Design of Axially Loaded Members

5.1 Introduction

Timber sections are commonly used in construction as *axially loaded members* or members in *combined axial force and bending*. Members of a truss, posts or columns, vertical wall studs and bracing elements are typical examples.

This chapter deals in detail with the general considerations necessary for the design of compression and tension members and describes the design details of solid section rectangular timber members. Design methods for glulam columns and columns of simple composites are described in Chapters 6 and 8 respectively.

5.2 Design of compression members

Compression members include posts or columns, vertical wall studs, and struts in trusses and girders. Permissible stresses for timber compression members are governed by the particular conditions of service and loading defined in Clauses 2.6.2, 2.8 and 2.9 of BS 5268 : Part 2 : 1996 which relate to different service class conditions, duration of loading and load-sharing systems respectively; and also by the additional factors given in Clause 2.11 of the code which are detailed here. Clause 2.11 deals with design of compression members and divides them into two categories:

(1) members subject to axial compression (without bending), and
(2) members subject to combined axial compression and bending (this may be due to applied eccentric compressive force).

5.2.1 *Design considerations*

The main design considerations for compression members are:

(1) Slenderness ratio. This relates to positional restraint of ends, lateral restraint along the length and cross-sectional dimensions of the member.
(2) Axial compression and bending stresses.

5.2.2 *Slenderness ratio, λ*

The load-carrying capacity of compression members is a function of the slenderness ratio, λ, which is calculated as the effective length, L_e divided by the radius of gyration, i:

$$\lambda = \frac{L_e}{i} \qquad (5.1)$$

The radius of gyration, i, is given by

$$i = \sqrt{\left(\frac{I}{A}\right)}$$

where I is the second moment of area and A is the cross-sectional area of the member. For rectangular sections, where b is the least lateral dimension (Fig. 5.1), the value of i simplifies to:

$$i = \frac{b}{\sqrt{12}} \qquad (5.2)$$

Clause 2.11.4 of BS 5268 : Part 2 : 1996 recommends that the slenderness ratio should not exceed a value of:

- $\lambda = 180$, for compression members carrying dead and imposed loads other than loads resulting from wind,
- $\lambda = 250$, for any members subject to reversal of axial stress solely from the effect of wind and any compression member carrying self-weight and wind loads only.

The effective length, L_e, of a column (given in Clause 2.11.3) should be derived from either:

(1) The deflected form of compression member as affected by any restraint and or fixing moment(s). Then the effective length is considered as the distance between adjacent points of zero bending moment, or

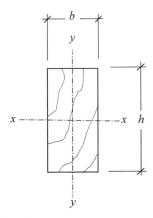

Fig. 5.1 Cross-section of a column.

Table 5.1 Effective length of compression members (Table 18, BS 5268 : Part 2)

End conditions	L_e/L
Restrained at both ends in position and in direction	0.7
Restrained at both ends in position and one end in direction	0.85
Restrained at both ends in position but not in direction	1.0
Restrained at one end in position and in direction and at the other end in direction but not in position	1.5
Restrained at one end in position and in direction and free at the other end	2.0

(2) Table 18 of the code for the particular end conditions at the column ends. Then the effective length is obtained by multiplying a relevant coefficient taken from this table by the actual length, L.

$$L_e = \text{coefficient} \times L \tag{5.3}$$

The end conditions defined in Table 18 of BS 5268 : Part 2 are reproduced here as Table 5.1 and are illustrated in Fig. 5.2.

For compression members with slenderness ratios equal to or greater than 5, Clause 2.11.5 of the code requires that the permissible compressive stresses are further modified by K_{12} the modification factor for compression members.

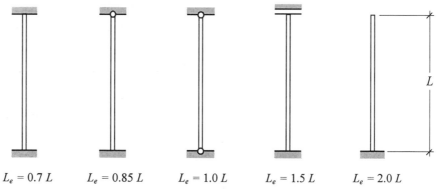

$L_e = 0.7\,L$ $L_e = 0.85\,L$ $L_e = 1.0\,L$ $L_e = 1.5\,L$ $L_e = 2.0\,L$

Fig. 5.2 Effective lengths and end conditions.

5.2.3 *Modification factor for compression members, K_{12}*

The modification factor for compression members, K_{12}, can be determined using either Table 19 of BS 5268 or calculated from the equation given in Annex B of the code. For either method, the minimum value of modulus of

Table 5.2 Modification factor K_{12} for compression members (Table 19, BS 5268 : Part 2)

Value of K_{12}

Value of slenderness ratio $\lambda(=L_e/i)$

Equivalent L_e/b (for rectangular sections)

$E/\sigma_{c,\parallel}$	<5	5	10	20	30	40	50	60	70	80	90	100	120	140	160	180	200	220	240	250
	<1.4	1.4	2.9	5.8	8.7	11.6	14.5	17.3	20.2	23.1	26.0	28.9	34.7	40.5	46.2	52.0	57.8	63.6	69.4	72.3
400	1.000	0.975	0.951	0.896	0.827	0.735	0.621	0.506	0.408	0.330	0.271	0.225	0.162	0.121	0.094	0.075	0.061	0.051	0.043	0.040
500	1.000	0.975	0.951	0.899	0.837	0.759	0.664	0.562	0.465	0.385	0.320	0.269	0.195	0.148	0.115	0.092	0.076	0.063	0.053	0.049
600	1.000	0.975	0.951	0.901	0.843	0.774	0.692	0.601	0.511	0.430	0.363	0.307	0.226	0.172	0.135	0.109	0.089	0.074	0.063	0.058
700	1.000	0.975	0.951	0.902	0.848	0.784	0.711	0.629	0.545	0.467	0.399	0.341	0.254	0.195	0.154	0.124	0.102	0.085	0.072	0.067
800	1.000	0.975	0.952	0.903	0.851	0.792	0.724	0.649	0.572	0.497	0.430	0.371	0.280	0.217	0.172	0.139	0.115	0.096	0.082	0.076
900	1.000	0.976	0.952	0.904	0.853	0.797	0.734	0.665	0.593	0.522	0.456	0.397	0.304	0.237	0.188	0.153	0.127	0.106	0.091	0.084
1000	1.000	0.976	0.952	0.904	0.855	0.801	0.742	0.677	0.609	0.542	0.478	0.420	0.325	0.255	0.204	0.167	0.138	0.116	0.099	0.092
1100	1.000	0.976	0.952	0.905	0.856	0.804	0.748	0.687	0.623	0.559	0.497	0.440	0.344	0.272	0.219	0.179	0.149	0.126	0.107	0.100
1200	1.000	0.976	0.952	0.905	0.857	0.807	0.753	0.695	0.634	0.573	0.513	0.457	0.362	0.288	0.233	0.192	0.160	0.135	0.116	0.107
1300	1.000	0.976	0.952	0.905	0.858	0.809	0.757	0.701	0.643	0.584	0.527	0.472	0.378	0.303	0.246	0.203	0.170	0.144	0.123	0.115
1400	1.000	0.976	0.952	0.906	0.859	0.811	0.760	0.707	0.651	0.595	0.539	0.486	0.392	0.317	0.259	0.214	0.180	0.153	0.131	0.122
1500	1.000	0.976	0.952	0.906	0.860	0.813	0.763	0.712	0.658	0.603	0.550	0.498	0.405	0.330	0.271	0.225	0.189	0.161	0.138	0.129
1600	1.000	0.976	0.952	0.906	0.861	0.814	0.766	0.716	0.664	0.611	0.559	0.508	0.417	0.342	0.282	0.235	0.198	0.169	0.145	0.135
1700	1.000	0.976	0.952	0.906	0.861	0.815	0.768	0.719	0.669	0.618	0.567	0.518	0.428	0.353	0.292	0.245	0.207	0.177	0.152	0.142
1800	1.000	0.976	0.952	0.906	0.862	0.816	0.770	0.722	0.673	0.624	0.574	0.526	0.438	0.363	0.302	0.254	0.215	0.184	0.159	0.148
1900	1.000	0.976	0.952	0.907	0.862	0.817	0.772	0.725	0.677	0.629	0.581	0.534	0.447	0.373	0.312	0.262	0.223	0.191	0.165	0.154
2000	1.000	0.976	0.952	0.907	0.863	0.818	0.773	0.728	0.681	0.634	0.587	0.541	0.455	0.382	0.320	0.271	0.230	0.198	0.172	0.160

elasticity E (i.e. E_{min}) should be used in all cases including when load sharing is present. The value of $\sigma_{c,//}$ for use in either method, should be the grade compressive stress, $\sigma_{c,g,//}$, (given in Tables 7–12a of the code) modified only for moisture content, duration of loading and size where appropriate. This will necessitate that all relevant load cases be considered, as the K_{12} value will differ with each load duration.

It is to be noted that for members comprising two or more pieces connected together in parallel and acting together to support the loads, the minimum modulus of elasticity should be modified by K_9 (see Table 17) or K_{28} (see Table 22) of the code. For horizontally laminated members, the modified mean modulus of elasticity should be used (see Clauses 3.2 and 3.6 of the code).

Table 19 of BS 5268 : Part 2 is reproduced here as Table 5.2 and the equation given in Annex B for calculation of K_{12} is in the following form:

$$K_{12} = \left\{ \frac{1}{2} + \frac{(1+\eta)\pi^2 E}{3\lambda^2 \sigma_c} \right\} - \left[\left\{ \frac{1}{2} + \frac{(1+\eta)\pi^2 E}{3\lambda^2 \sigma_c} \right\}^2 - \frac{\pi^2 E}{1.5\lambda^2 \sigma_c} \right]^{1/2} \quad (5.4)$$

where:
σ_c = permissible stress for a very short column ($\lambda < 5$), $\sigma_{c,g,//} \times K_2 K_3$
E = minimum modulus of elasticity, E_{min}

λ = slenderness ratio, $\dfrac{L_e}{i}$

$\eta = 0.005\,\lambda$

5.2.4   *Members subjected to axial compression only (Clause 2.11.5)*

An axially loaded column has its line of action of load passing through the centroidal axis of the column, see Fig. 5.3(a).

(1)   The *axial compressive stress*, $\sigma_{c,a,//}$, is given by:

$$\sigma_{c,a,//} = \frac{P}{A} \quad (5.5)$$

where P is the axial compressive load and A the cross-sectional area.

(2)   The *permissible compressive stress*, $\sigma_{c,adm,//}$, is calculated as the product of grade compressive stress parallel to grain, $\sigma_{c,g,//}$, and any relevant modification factors (K-factors). These are K_2 for wet exposure condition (if applicable), K_3 for load-duration, K_8 for load-sharing systems (if applicable) and the modification factor for compression members, K_{12}, if appropriate:

if $\lambda < 5$ $\sigma_{c,adm,//} = \sigma_{c,g,//} \times K_2 K_3 K_8$ (5.6a)

and for $\lambda \geq 5$ $\sigma_{c,adm,//} = \sigma_{c,g,//} \times K_2 K_3 K_8 K_{12}$ (5.6b)

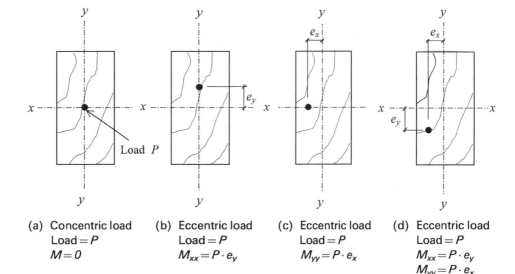

(a) Concentric load
 Load $=P$
 $M=0$

(b) Eccentric load
 Load $=P$
 $M_{xx}=P \cdot e_y$

(c) Eccentric load
 Load $=P$
 $M_{yy}=P \cdot e_x$

(d) Eccentric load
 Load $=P$
 $M_{xx}=P \cdot e_y$
 $M_{yy}=P \cdot e_x$

Fig. 5.3 Axial concentric and eccentric loads.

The compression member is designed so that the applied compressive stress, $\sigma_{c,a,/\!/}$, does not exceed the permissible compressive stress parallel to the grain, $\sigma_{c,adm,/\!/}$

$$\sigma_{c,a,/\!/} \le \sigma_{c,adm,/\!/}$$

5.2.5 *Members subjected to axial compression and bending (Clause 2.11.6)*

This includes compression members subject to eccentric loading, where load acts through a point at a certain distance from the centroidal axis, which can be equated to an axial compression force and bending moment, see Fig. 5.3(b), (c) or (d). Members which are restrained at both ends in position but not direction, which covers most real situations, should be so proportioned that:

$$\frac{\sigma_{m,a,/\!/}}{\sigma_{m,adm,/\!/}\left(1 - \frac{1.5\sigma_{c,a,/\!/}}{\sigma_e} K_{12}\right)} + \frac{\sigma_{c,a,/\!/}}{\sigma_{c,adm,/\!/}} \le 1 \tag{5.7}$$

where:

$$\sigma_{m,a,/\!/} = \text{applied bending stress} = \frac{M}{Z}$$

$$\sigma_{m,adm,/\!/} = \text{permissible bending stress} = \sigma_{m,g,/\!/} \times K_2 K_3 K_6 K_7 K_8$$

$$\sigma_{c,a,/\!/} = \text{applied compression stress} = \frac{P}{A}$$

$$\sigma_{c,adm,/\!/} = \text{permissible compression stress} = \sigma_{c,g,/\!/} \times K_2 K_3 K_8 K_{12}$$

$$\sigma_e = \text{Euler critical stress} = \frac{\pi^2 EI}{(L_e/i)^2} \text{ and } E = E_{min}.$$

Equation (5.7) is the interaction formula used to ensure that lateral instability does not arise in compression members subject to axial force and bending. Thus if the column is subject to compressive loading only, then $\sigma_{m,a,/\!/} = 0$, and hence the equation simplifies to $\sigma_{c,a,/\!/}/\sigma_{c,adm,/\!/} \leq 1$. Alternatively, if the column is subject to bending only, i.e. $\sigma_{c,a,/\!/} = 0$, then the equation simplifies to $\sigma_{m,a,/\!/}/\sigma_{m,adm,/\!/} \leq 1$.

5.2.6 *Design of load-bearing stud walls*

Stud walls are often constructed as load-bearing walls in timber framed housing. Details of a typical stud wall are shown in Fig. 5.4.

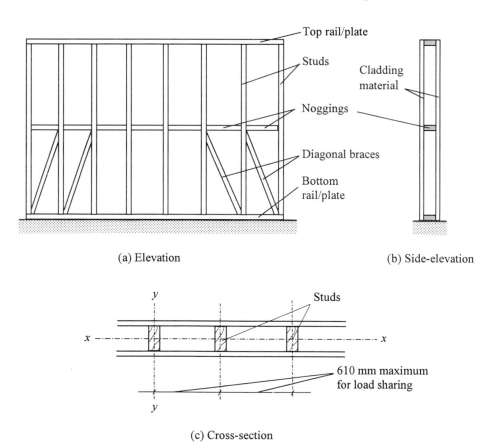

(a) Elevation   (b) Side-elevation

(c) Cross-section

Fig. 5.4   Details of a typical stud wall.

Stud walls consist of vertical timber members, commonly referred to as studs, which are held in position by nailing them to timber rails or plates, located along the top and bottom ends of the studs. These walls can be designed to resist both vertical and lateral loadings; wind load being a typical example of a lateral load. Each stud may be considered to be laterally restrained about its $y-y$ axis, fully or partially (for example, at mid height), either by the cladding/sheathing material, such as plasterboard and by internal noggings or diagonal bracing. Therefore in situations where the cladding/sheathing material is properly attached to the stud along its whole length, the strength of the stud can be calculated about its $x-x$ axis; otherwise the greater of the slenderness ratios about the individual stud's $x-x$ and $y-y$ axes should be considered in the design calculations.

The following general considerations may apply in designing stud walls:

(1) Studs are considered as a series of posts (columns) carrying concentric (axial compression) or combined bending and axial compressive loads.
(2) *Load-sharing*: If stud spacing is less than 610 mm (which it usually is), the load-sharing factor ($K_8 = 1.1$) applies.
(3) *End restraints*: In general the individual studs are assumed to be laterally restrained in position but not in direction (pinned ends), as these walls are normally provided with a top and bottom rail. In other situations, for example where the studs' bases are cast in concrete, the appropriate end conditions should be used (see Table 5.1).

5.3   Design of tension members (Clause 2.12)

Permissible stresses for timber tension members are governed by the particular conditions of service and loading defined in Clauses 2.6.2, 2.8 and 2.9 of BS 5268 : Part 2 which relate to different service class conditions, duration of loading and load-sharing systems respectively; and also by the additional factors given in Clause 2.12 of the code which are detailed here.

With tension members, since there is no tendency to buckle, the ratio of length to thickness (i.e. the slenderness ratio, λ) is not critical. In truss frameworks sometimes a member may undergo compression for a short duration due to wind loading and hence it is recommended that the slenderness ratio, λ, for truss members is limited to 250.

The design of tension members involves trial and error. The capacity of a tension member at its weakest point, for example at connection points, should be determined. Generally, the cross-sectional dimensions are found first by assuming a connector arrangement. After connectors are designed, the strength of the member is again checked based on its net cross-sectional area at the point of connection.

5.3.1   *Design considerations*

The main design considerations for tension members are:

(1)   Tensile stress for members subjected to axial loading only.
(2)   Combined bending and tensile stresses for members subjected to lateral loading as well as the axial tension.

In both cases, as before, the permissible stress value is calculated as the product of the grade stress and the appropriate modification factors for a particular service and loading conditions, and is compared with the applied stress in the member, where:

Permissible stress ($=$ grade stress $\times$ K-factors) $\geq$ applied stress

BS 5268 : Part 2 recommends that in the calculation of the permissible tensile stress value the effect of the width of the timber section K_{14} should be considered.

5.3.2   *Width factor, K_{14}*

The grade tension stresses given in Tables 7–12a of BS 5268 : Part 2 apply to members having a width (i.e. the greater transverse dimension), h, of 300 mm (see Fig. 5.1). For other widths of members assigned to a strength class, the grade tension stresses should be multiplied by the width modification factor, K_{14}, where:

for $h \leq 72$ mm $K_{14} = 1.17$

and for $h > 72$ mm $K_{14} = \left(\dfrac{300}{h}\right)^{0.11}$

5.3.3   *Members subjected to axial tension only*

An axially loaded column has its line of action of load passing through the centroidal axis of the column.

(1)   The *applied tensile stress*, $\sigma_{t,a,//}$, in an axially loaded timber member is calculated from the following equation:

$$\sigma_{t,a,//} = \frac{T}{A_{net}} \tag{5.8}$$

where:
$T =$ tensile force
$A_{net} =$ net cross-sectional area.

(2) The *permissible tensile stress*, $\sigma_{t,adm,\parallel}$, is calculated as the product of the grade tensile stress, $\sigma_{t,g,\parallel}$, and any relevant modification factors (*K*-factors) as follows:

$$\sigma_{t,adm,\parallel} = \sigma_{t,g,\parallel} \times K_2 K_3 K_8 K_{14} \tag{5.9}$$

where K_2, K_3 and K_8 are general modification factors for service class 3, load-duration and load-sharing systems respectively, which were described in detail in Chapter 2. K_{14} is the *width factor* as described earlier.

In general, the value of applied tensile stress, $\sigma_{t,a,\parallel}$, should not exceed the permissible tensile stress, $\sigma_{t,adm,\parallel}$, hence:

$$\sigma_{t,a,\parallel} \leq \sigma_{t,adm,\parallel}$$

5.3.4 *Combined bending and tensile stresses*

In members that are subjected to lateral loading as well as the axial tension, the position of maximum stress occurs at the point of maximum bending moment (Fig. 5.5). Clause 2.12.3 of the code requires that the sum of the ratios of the applied tensile and bending stresses to those of the permissible ones (i.e. interaction quantity) must not exceed unity:

$$\frac{\sigma_{m,a,\parallel}}{\sigma_{m,adm,\parallel}} + \frac{\sigma_{t,a,\parallel}}{\sigma_{t,adm,\parallel}} \leq 1 \tag{5.10}$$

where:

applied tensile stress, $\sigma_{t,a,\parallel} = \dfrac{T}{A_{net}}$

permissible tensile stress, $\sigma_{t,adm,\parallel} = \sigma_{t,g,\parallel} \times K_2 K_3 K_8 K_{14}$

applied bending stress, $\sigma_{m,a,\parallel} = \dfrac{M \cdot y}{I} = \dfrac{M}{Z}$

permissible bending stress, $\sigma_{m,adm,\parallel} = \sigma_{m,g,\parallel} \times K_2 K_3 K_6 K_7 K_8$

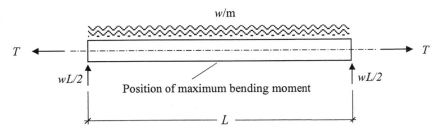

Fig. 5.5 Tension member subjected to lateral loading.

5.4   Design examples

Example 5.1   Load capacity of a timber column

A timber column in strength class C18 is 4 m in height with a rectangular cross-section of 97 mm × 145 mm as shown in Fig. 5.6. The column is restrained at both ends in position but not in direction and is subjected to service class 2 conditions.
(a)   Determine the maximum axial long-term load that the column can support.
(b)   Check that the column is adequate to resist a long-term axial load of 12 kN and a bending momentof 0.8 kNm about its *x–x* axis.

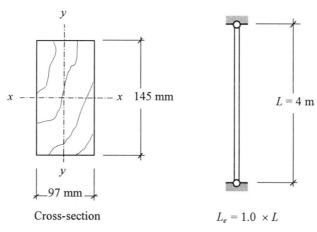

Cross-section $L_e = 1.0 \times L$

Fig. 5.6   Column details (Example 5.1).

Definitions

Force, kN	N := newton
Length, m	kN := $10^3 \cdot$ N
Cross-sectional dimensions, mm	Direction parallel to grain, //
Stress, Nmm^{-2}	Direction perpendicular to grain, *pp*

1.   Geometrical properties

BS 5268 : Part 2, Table 18

Column length, *L* $L := 4.0 \cdot m$

Effective length, L_e $L_e := 1.0 \cdot L$

$\qquad\qquad\qquad\qquad\qquad L_e = 4 \circ m$

Width of section, *b* $b := 97 \cdot mm$

Depth of section, *h* $h := 145 \cdot mm$

Cross-sectional area, *A* $A := b \cdot h$

$\qquad\qquad\qquad\qquad\qquad A = 14065 \circ mm^2$

Second moment of area, I_{xx} $I_{xx} := \frac{1}{12} \cdot b \cdot h^3$

$\qquad\qquad\qquad\qquad\qquad I_{xx} = 2.46 \times 10^7 \circ mm^4$

Second moment of area, I_{yy}	$I_{yy} := \frac{1}{12} \cdot h \cdot b^3$
	$I_{yy} = 1.1 \times 10^7 \circ \text{mm}^4$

For a rectangular section

Radius of gyration, i_{yy}	$i_{yy} := \dfrac{b}{\sqrt{12}}$
	$i_{yy} = 28 \circ \text{mm}$
Section modulus, Z_{xx}	$Z_{xx} := \dfrac{b \cdot h^2}{6}$
	$Z_{xx} = 339\,904.17 \circ \text{mm}^3$

2. Check slenderness ratio, λ

$$\lambda := \frac{L_e}{i_{yy}}$$
$$\lambda = 142.85$$
$$< 180, \quad \text{satisfactory}$$

3. Grade stresses

BS 5268: :Part 2, Table 7
Timber strength class, C18

Bending parallel to grain	$\sigma_{m.g.\!/\!/} := 5.8 \cdot \text{N} \cdot \text{mm}^{-2}$
Compression parallel to grain	$\sigma_{c.g.\!/\!/} := 7.1 \cdot \text{N} \cdot \text{mm}^{-2}$
Minimum modulus of elasticity	$E_{min} := 6000 \cdot \text{N} \cdot \text{mm}^{-2}$

4. K-factors

Service class 2 (K_2, Table 13)	$K_2 := 1$
Load duration (K_3, Table 14)	$K_3 := 1$ for long-term
Form factor (K_6, Clause 2.10.5)	$K_6 := 1$
Depth factor (K_7, Clause 2.10.6)	$K_7 := \left(\dfrac{300 \cdot \text{mm}}{h} \right)^{0.11}$
	$K_7 = 1.08$
No load sharing	$K_8 := 1$
(K_8, Clauses 2.9 and 2.11.5)	

The modification factor for compression member, K_{12}, can be calculated using either one of the following methods:

(1)   Using equation in Annex B:

$$\sigma_c := \sigma_{c.g.\!/\!/} \cdot K_2 \cdot K_3 \qquad \eta := 0.005 \cdot \lambda \qquad E := E_{min}$$

$$K_{12} := \frac{1}{2} + \frac{(1+\eta) \cdot \pi^2 \cdot E}{3 \cdot \lambda^2 \cdot \sigma_c} - \sqrt{\left[\frac{1}{2} + \frac{(1+\eta) \cdot \pi^2 \cdot E}{3 \cdot \lambda^2 \cdot \sigma_c} \right]^2 - \frac{\pi^2 \cdot E}{1.5 \cdot \lambda^2 \cdot \sigma_c}}$$

$$K_{12} = 0.22$$

(2) Using Table 19:

For $\dfrac{E_{min}}{\sigma_{c.g.\!/\!/} \cdot K_2 \cdot K_3} = 845.07$ and $\lambda = 142.85$ thus, $K_{12} = 0.22$

Part (a)

5. *Permissible compressive stress*

$$\sigma_{c.adm.\!/\!/} := \sigma_{c.g.\!/\!/} \cdot K_2 \cdot K_3 \cdot K_8 \cdot K_{12}$$
$$\sigma_{c.adm.\!/\!/} = 1.55 \circ \text{N} \cdot \text{mm}^{-2}$$

Hence the axial long-term load capacity of the column is calculated by rearranging, $\sigma = P/A$. Thus

$$P_{capacity} := \sigma_{c.adm.\!/\!/} \cdot A$$
$$P_{capacity} = 21.79 \circ \text{kN}$$

Part (b)

6. *Loading*

Axial load $P := 12 \cdot \text{kN}$

Applied moment $M := 0.8 \cdot \text{kN} \cdot \text{m}$

7. *Compression and bending stresses*

Applied compressive stress $\sigma_{c.a.\!/\!/} := \dfrac{P}{A}$

$\sigma_{c.a.\!/\!/} = 0.85 \circ \text{N} \cdot \text{mm}^{-2}$

Permissible compressive stress $\sigma_{c.adm.\!/\!/} := \sigma_{c.g.\!/\!/} \cdot K_2 \cdot K_3 \cdot K_8 \cdot K_{12}$

$\sigma_{c.adm.\!/\!/} = 1.55 \circ \text{N} \cdot \text{mm}^{-2}$

Compression satisfactory

Applied bending stress $\sigma_{m.a.\!/\!/} := \dfrac{M}{Z_{xx}}$

$\sigma_{m.a.\!/\!/} = 2.35 \circ \text{N} \cdot \text{mm}^{-2}$

Permissible bending stress $\sigma_{m.adm.\!/\!/} := \sigma_{m.g.\!/\!/} \cdot K_2 \cdot K_3 \cdot K_6 \cdot K_7 \cdot K_8$

$\sigma_{m.adm.\!/\!/} = 6.28 \circ \text{Nmm}^{-2}$

Check interaction quantity

Euler critical stress $\sigma_e := \dfrac{\pi^2 \cdot E_{min}}{\lambda^2}$

$\sigma_e = 2.9 \circ \text{N} \cdot \text{mm}^{-2}$

$$\dfrac{\sigma_{m.a.\!/\!/}}{\sigma_{m.adm.\!/\!/} \cdot \left(1 - \dfrac{1.5 \cdot \sigma_{c.a.\!/\!/}}{\sigma_e} \cdot K_{12}\right)} + \dfrac{\sigma_{c.a.\!/\!/}}{\sigma_{c.adm.\!/\!/}} = 0.97 < 1$$

Interaction quantity is satisfactory

Therefore a 97 mm × 145 mm timber section in strength class C18 is satisfactory

Example 5.2   Design of an eccentrically loaded column

For the design data given below, check that a 100 mm × 250 mm rough sawn section, as shown in Fig. 5.7, is adequate as a column if the load is applied 90 mm eccentric to its x–x axis. The column is 3.75 m in height and has its ends restrained in position but not in direction.

Design data:

Timber	Whitewood grade SS
Service class	2
Design load (long-term)	25 kN
Design load (medium-term)	30 kN

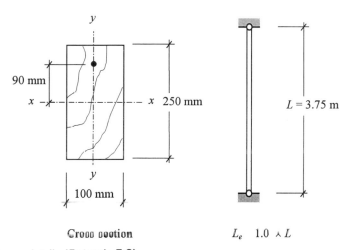

Fig. 5.7   Column details (Example 5.2).

Definitions

Force, kN	N := newton
Length, m	kN := $10^3 \cdot$ N
Cross-sectional dimensions, mm	Direction parallel to grain, //
Stress, Nmm^{-2}	Direction perpendicular to grain, *pp*

1.   Geometrical properties

BS 5268 : Part 2, Table 18

Column length, L $L := 3.75 \cdot$ m

Effective length, L_e $L_e := 1.0 \cdot L$

                                  $L_e = 3.75 \circ$ m

Width of section b $b := 100 \cdot$ mm

Depth of section, h $h := 250 \cdot$ mm

Cross-sectional area, A

$A := b \cdot h$
$A = 25\,000 \circ mm^2$

Second moment of area, I_{xx}

$I_{xx} := \frac{1}{12} \cdot b \cdot h^3$
$I_{xx} = 1.3 \times 10^8 \circ mm^4$

Second moment of area, I_{yy}

$I_{yy} := \frac{1}{12} \cdot h \cdot b^3$
$I_{yy} = 2.08 \times 10^7 \circ mm^4$

For a rectangular section, radius of gyration: i_{yy}

$i_{yy} := \dfrac{b}{\sqrt{12}}$

$i_{yy} = 28.87 \circ mm$

Section modulus, Z_{xx}

$Z_{xx} := \dfrac{b \cdot h^2}{6}$

$Z_{xx} = 1.04 \times 10^6 \circ mm^3$

2.   Check slenderness ratio, λ

$\lambda := \dfrac{L_e}{i_{yy}}$

$\lambda = 129.9$
<180, satisfactory

3.   Grade stresses

BS 5268 : Part 2, Tables 2 and 7
Strength classification
Bending parallel to grain
Compression parallel to grain
Minimum modulus of elasticity

Whitewood grade SS $= C24$
$\sigma_{m.g.//} := 7.5 \cdot N \cdot mm^{-2}$
$\sigma_{c.g.//} := 7.9 \cdot N \cdot mm^{-2}$
$E_{min} := 7200 \cdot N \cdot mm^{-2}$

4.   K-factors

Service class 2 (K_2, Table 13)
Load duration (K_3, Table 14)
   for long- and medium-terms
Form factor (K_6, Clause 2.10.5)

$K_2 := 1$
$K_{3.long} := 1$ and $K_{3.med} := 1.25$

$K_6 := 1$

Depth factor (K_7, Clause 2.10.6)

$K_7 := \left(\dfrac{300 \cdot mm}{h} \right)^{0.11}$

$K_7 = 1.02$

No load sharing
   (K_8, Clauses 2.9 & 2.11.5)

$K_8 := 1$

Modification factor for compression member, K_{12}, can be calculated using either one of the following methods:

(1) Using equation in Annex B:

  (a) Long-term loading: $\sigma_c := \sigma_{c.g.\parallel} \cdot K_2 \cdot K_{3.long}$ $\eta := 0.005 \cdot \lambda$ $E := E_{min}$

$$K_{12.long} := \frac{1}{2} + \frac{(1+\eta) \cdot \pi^2 \cdot E}{3 \cdot \lambda^2 \cdot \sigma_c} - \sqrt{\left(\frac{1}{2} + \frac{(1+\eta) \cdot \pi^2 \cdot E}{3 \cdot \lambda^2 \cdot \sigma_c}\right)^2 - \frac{\pi^2 \cdot E}{1.5 \cdot \lambda^2 \cdot \sigma_c}}$$

$$K_{12.long} = 0.27$$

  (b) Medium-term loading: $\sigma_c := \sigma_{c.g.\parallel} \cdot K_2 \cdot K_{3.med}$ $\eta := 0.005 \cdot \lambda$

$$E := E_{min}$$

$$K_{12.med} := \frac{1}{2} + \frac{(1+\eta) \cdot \pi^2 \cdot E}{3 \cdot \lambda^2 \cdot \sigma_c} - \sqrt{\left(\frac{1}{2} + \frac{(1+\eta) \cdot \pi^2 \cdot E}{3 \cdot \lambda^2 \cdot \sigma_c}\right)^2 - \frac{\pi^2 \cdot E}{1.5 \cdot \lambda^2 \cdot \sigma_c}}$$

$$K_{12.med} = 0.23$$

(2) Using Table 19:

  (a) Long-term loading:

$$\frac{E_{min}}{\sigma_{c.g.\parallel} \cdot K_2 \cdot K_{3.long}} = 911.39 \quad \text{and} \quad \lambda = 129.9 \quad \text{thus,} \quad K_{12.long} = 0.27$$

  (b) Medium-term loading:

$$\frac{E_{min}}{\sigma_{c.g.\parallel} \cdot K_2 \cdot K_{3.med}} = 729.11 \quad \text{and} \quad \lambda - 129.9 \quad \text{thus,} \quad K_{12.med} = 0.23$$

5. *Loading*

Eccentricity, e_y $e_y := 90 \cdot mm$
(a) Long-term loading:
Axial load $P_{long} := 25 \cdot kN$
Moment = axial load × eccentricity $M_{long} := P_{long} \cdot e_y$
                                         $M_{long} = 2.25 \circ kN \cdot m$

(b) Medium-term loading:
Axial load $P_{med} := 30 \cdot kN$
Moment = axial load × eccentricity $M_{med} := P_{med} \cdot e_y$
                                         $M_{med} = 2.7 \circ kN \cdot m$

6. *Compression and bending stress*

(a) Long-term loading: $K_3 := K_{3.long}$ $K_{12} := K_{12.long}$ $P := P_{long}$ $M := M_{long}$

Applied compressive stress $\sigma_{c.a.\parallel} := \frac{P}{A}$

                                         $\sigma_{c.a.\parallel} = 1 \circ N \cdot mm^{-2}$
Permissible compressive stress $\sigma_{c.adm.\parallel} := \sigma_{c.g.\parallel} \cdot K_2 \cdot K_3 \cdot K_8 \cdot K_{12}$
                                         $\sigma_{c.adm.\parallel} = 2.13 \circ N \cdot mm^{-2}$
                                         Compression satisfactory

Applied bending stress

$$\sigma_{m.a.\parallel} := \frac{M}{Z_{xx}}$$

$$\sigma_{m.a.\parallel} = 2.16 \circ \mathrm{N} \cdot \mathrm{mm}^{-2}$$

Permissible bending stress

$$\sigma_{m.adm.\parallel} := \sigma_{m.g.\parallel} \cdot K_2 \cdot K_3 \cdot K_6 \cdot K_7 \cdot K_8$$

$$\sigma_{m.adm.\parallel} = 7.65 \circ \mathrm{N} \cdot \mathrm{mm}^{-2}$$

Bending satisfactory

Check interaction quantity

Euler critical stress

$$\sigma_e := \frac{\pi^2 \cdot E_{min}}{\lambda^2}$$

$$\sigma_e = 4.21 \circ \mathrm{N} \cdot \mathrm{mm}^{-2}$$

$$\frac{\sigma_{m.a.\parallel}}{\sigma_{m.adm.\parallel} \cdot \left(1 - \dfrac{1.5 \cdot \sigma_{c.a.\parallel}}{\sigma_e} \cdot K_{12}\right)} + \frac{\sigma_{c.a.\parallel}}{\sigma_{c.adm.\parallel}} = 0.78 < 1$$

Interaction quantity satisfactory

(b)   Medium term loading: $K_3 := K_{3.med}$ $K_{12} := K_{12.med}$ $P := P_{med}$ $M := M_{med}$

Applied compressive stress

$$\sigma_{c.a.\parallel} := \frac{P}{A}$$

$$\sigma_{c.a.\parallel} = 1.2 \circ \mathrm{N} \cdot \mathrm{mm}^{-2}$$

Permissible compressive stress

$$\sigma_{c.adm.\parallel} := \sigma_{c.g.\parallel} \cdot K_2 \cdot K_3 \cdot K_8 \cdot K_{12}$$

$$\sigma_{c.adm.\parallel} = 2.26 \circ \mathrm{N} \cdot \mathrm{mm}^{-2}$$

Compression satisfactory

Applied bending stress

$$\sigma_{m.a.\parallel} := \frac{M}{Z_{xx}}$$

$$\sigma_{m.a.\parallel} = 2.59 \circ \mathrm{N} \cdot \mathrm{mm}^{-2}$$

Permissible bending stress

$$\sigma_{m.adm.\parallel} := \sigma_{m.g.\parallel} \cdot K_2 \cdot K_3 \cdot K_6 \cdot K_7 \cdot K_8$$

$$\sigma_{m.adm.\parallel} = 9.56 \circ \mathrm{N} \cdot \mathrm{mm}^{-2}$$

Bending satisfactory

Check interaction quantity

Euler critical stress

$$\sigma_e := \frac{\pi^2 \cdot E_{min}}{\lambda^2}$$

$$\sigma_e = 4.21 \circ \mathrm{N} \cdot \mathrm{mm}^{-2}$$

$$\frac{\sigma_{m.a.\parallel}}{\sigma_{m.adm.\parallel} \cdot \left(1 - \dfrac{1.5 \cdot \sigma_{c.a.\parallel}}{\sigma_e} \cdot K_{12}\right)} + \frac{\sigma_{c.a.\parallel}}{\sigma_{c.adm.\parallel}} = 0.83 < 1$$

Interaction quantity satisfactory

Therefore a 100 mm × 250 mm timber section in strength class C24 is satisfactory

Example 5.3   Design of a load-bearing stud wall

A stud wall, as shown in Fig. 5.8, has an overall height of 4.0 m. the vertical studs are positioned at 600 mm centres with noggings at mid-height. Carry out design calculations to show that studs of 44 mm × 100 mm section in C16 timber under service class 1 are suitable to sustain a long-duration load of 10.0 kN/m.

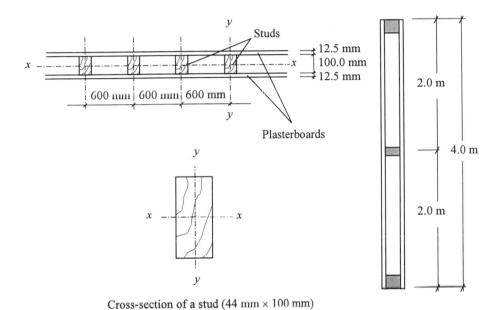

Cross-section of a stud (44 mm × 100 mm)

Fig. 5.8   Stud wall details (Example 5.3).

Definitions

Force, kN	$N := \text{newton}$
Length, m	$kN := 10^3 \cdot N$
Cross-sectional dimensions, mm	Direction parallel to grain, //
Stress, Nmm^{-2}	Direction perpendicular to grain, *pp*

1.   Geometrical properties

BS 5268 : Part 2, Table 18
For each stud acting as a column, restrained about its *y–y* axis by plasterboards and restrained at both ends in position but not in direction:

Stud length, about *x–x* axis, L_x $L_x := 4.0 \cdot m$
Effective length, about $L_{e.x} := 1.0 \cdot L_x$
    x–x axis, $L_{e.z}$ $L_{e.x} = 4 \circ m$

Width of section, b	$b := 44 \cdot \text{mm}$
Depth of section, h	$h := 100 \cdot \text{mm}$
Cross-sectional area A	$A := b \cdot h$
	$A = 4400 \circ \text{mm}^2$
Second moment of area, I_{xx}	$I_{xx} := \frac{1}{12} \cdot b \cdot h^3$
	$I_{xx} = 3.67 \times 10^6 \circ \text{mm}^4$
For a rectangular section, radius of gyration, i_{xx}	$i_{xx} = \dfrac{h}{\sqrt{12}}$
	$i_{xx} = 28.87 \circ \text{mm}$

2. Check slenderness ratio, λ

about x–x axis

$$\lambda_{xx} := \frac{L_{e.x}}{i_{xx}}$$

$$\lambda_{xx} = 138.56$$

< 180, satisfactory

3. Grade stresses

BS 5268 : Part 2, Table 7
Timber strength class, C16
Bending parallel to grain $\sigma_{m.g.//} := 5.3 \cdot \text{N} \cdot \text{mm}^{-2}$
Compression parallel to grain $\sigma_{c.g.//} := 6.8 \cdot \text{N} \cdot \text{mm}^{-2}$
Minimum modulus of elasticity $E_{min} := 5800 \cdot \text{N} \cdot \text{mm}^{-2}$

4. K-factors

Service class 2 (K_2, Table 13)	$K_2 := 1$
Load duration (K_3, Table 14)	$K_3 := 1$
Form factor (K_6, Clause 2.10.5)	$K_6 := 1$
Load sharing applies (K_8, Clause 2.9)	$K_8 := 1.1$

Modification factor for compression member, K_{12}, can be calculated using either one of the following methods:

(1) Using equation in Annex B:

$$\sigma_c := \sigma_{c.g.//} \cdot K_2 \cdot K_3 \quad \lambda := \lambda_{xx} \quad \eta := 0.005 \cdot \lambda \quad E := E_{min}$$

$$K_{12} := \frac{1}{2} + \frac{(1+\eta) \cdot \pi^2 \cdot E}{3 \cdot \lambda^2 \cdot \sigma_c} - \sqrt{\left(\frac{1}{2} + \frac{(1+\eta) \cdot \pi^2 \cdot E}{3 \cdot \lambda^2 \cdot \sigma_c}\right)^2 - \frac{\pi^2 \cdot E}{1.5 \cdot \lambda^2 \sigma_c}}$$

$$K_{12} = 0.23$$

(2)   Using Table 19:

For $\dfrac{E_{min}}{\sigma_{c.g.\!/\!/} \cdot K_2 \cdot K_3} = 852.94$ and $\lambda = 138.56$ thus, $K_{12} = 0.23$

5.   Loading

Uniformly distributed axial load	$P := 10 \cdot \text{kN} \cdot \text{m}^{-1}$
Stud spacing, S_{stud}	$S_{stud} := 0.6 \cdot \text{m}$
Axial load per stud	$P_{stud} := P \cdot S_{stud}$
	$P_{stud} = 6 \circ \text{kN}$

6.   Compression stress

Applied compressive stress $\sigma_{c.a.\!/\!/} := \dfrac{P_{stud}}{A}$

$\sigma_{c.a.\!/\!/} = 1.36 \circ \text{N} \cdot \text{mm}^{-2}$

Permissible compressive stress $\sigma_{c.adm.\!/\!/} := \sigma_{c.g.\!/\!/} \cdot K_2 \cdot K_3 \cdot K_8 \cdot K_{12}$

$\sigma_{c.adm.\!/\!/} = 1.73 \circ \text{N} \cdot \text{mm}^{-2}$

Compression satisfactory

Therefore 44 mm × 100 mm timber sections in strength class C16 are satisfactory

Example 5.4   Design of a load-bearing stud wall with lateral load

A stud wall forming part of an open farm building, as shown in Fig. 5.9, has an overall height of 3.6 m. The vertical studs are built-in to concrete at the base, support a roof

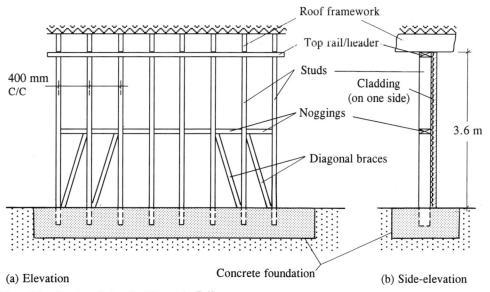

(a) Elevation

(b) Side-elevation

Fig. 5.9   Stud wall details (Example 5.4).

structure at the top, and are positioned at 400 mm centres with noggings at mid-height. The cladding materials are assumed to make no contribution to the structural stability of the framed structure. Carry out design calculations to show that studs of 47 mm × 100 mm section in C22 timber under service class 3 are suitable to sustain a medium-duration load of 5.4 kN/m and a wind loading of 0.975 kN/m² (3 s gust).

Definitions

Force, kN	$N := $ newton
Length, m	$kN := 10^3 \cdot N$
Cross-sectional dimensions, mm	Direction parallel to grain, //
Stress, Nmm^{-2}	Direction perpendicular to grain, *pp*

1. Geometrical properties

BS 5268 : Part 2, Table 18

Stud length, L $L := 3.6 \cdot m$

Stud length, about x–x axis, L_x $L_x := 3.6 \cdot m$

Stud length, about y–y axis, L_y $L_y := 1.8 \cdot m$

Effective length, about
     x–x axis, $L_{e.x}$ $L_{e.x} := 0.85 \cdot L_x$
$L_{e.x} = 3.06 \circ m$

Effective length, about
     y–y axis, $L_{e.y}$ is the greater of: $L_{e.y} := 0.85 \cdot L_y$ or $L_{e.y} := 1.0 \cdot L_y$

Thus $L_{e.y} := 1.0 \cdot L_y$
$L_{e.y} = 1.8 \circ m$

Width of section, b $b := 47 \cdot mm$

Depth of section, h $h := 100 \cdot mm$

Cross-sectional area, A $A := b \cdot h$
$A := 4.7 \times 10^3 \circ mm^2$

Second moment of area, I_{xx} $I_{xx} := \frac{1}{12} \cdot b \cdot h^3$
$I_{xx} = 3.92 \times 10^6 \circ mm^4$

Second moment of area, I_{yy} $I_{yy} := \frac{1}{12} \cdot h \cdot b^3$
$I_{yy} = 8.65 \times 10^5 \circ mm^4$

For a rectangular section,

     radius of gyration, i_{yy} $i_{yy} := \dfrac{b}{\sqrt{12}}$
$i_{yy} = 13.57 \circ mm$

     radius of gyration, i_{xx} $i_{xx} := \dfrac{h}{\sqrt{12}}$
$i_{xx} = 28.87 \circ mm$

Section modulus, Z_{xx} $Z_{xx} := \dfrac{b \cdot h^2}{6}$
$Z_{xx} = 7.83 \times 10^4 \circ mm^3$

2. *Check slenderness ratio,* λ

about y–y axis

$$\lambda_{yy} := \frac{L_{e.y}}{i_{yy}}$$

$$\lambda_{yy} = 132.67$$

about x–x axis

$$\lambda_{xx} := \frac{L_{e.x}}{i_{xx}}$$

$$\lambda_{xx} = 106$$

Both < 180, satisfactory

λ_{yy} is the critical one

3. *Grade stresses*

BS 5268 : Part 2, Table 7
Timber strength class, C22
Bending parallel to grain $\qquad \sigma_{m.g.\text{/\!/}} := 6.8 \cdot N \cdot mm^{-2}$
Compression parallel to grain $\quad \sigma_{c.g.\text{/\!/}} := 7.5 \cdot N \cdot mm^{-2}$
Minimum modulus of elasticity $\quad E_{min} := 6500 \cdot N \cdot mm^{-2}$

4. *K-factors*

Service class 3 (K_2, Table 13) $\qquad K_{2.bending} := 0.8 \quad K_{2.comp} := 0.6 \quad K_{2.E} := 0.8$
Load duration (K_3, Table 14) $\qquad K_{3.med} := 1.25 \quad$ (from roof)
$\qquad\qquad\qquad\qquad\qquad\qquad K_{3.v.short} := 1.75 \quad$ (from wind)

Form factor (K_6, Clause 2.10.5) $\quad K_6 := 1$

Depth factor (K_7, Clause 2.10.6) $\quad K_7 := \left(\dfrac{300 \cdot mm}{h} \right)^{0.11}$

$$K_7 = 1.13$$

Load sharing applies $\qquad\qquad K_8 := 1.1$
  (K_8, Clause 2.9)

Modification factor for compression member, K_{12}

(1) Medium-term loading (no wind):
(a) Using equation in Annex B:

$$K_3 := K_{3.med} \qquad \sigma_c := \sigma_{c.g.\text{/\!/}} \cdot K_{2.comp} \cdot K_3$$

$$\lambda := \lambda_{yy} \qquad \eta := 0.005 \cdot \lambda \qquad E := E_{min} \cdot K_{2.E}$$

$$K_{12.med} := \frac{1}{2} + \frac{(1+\eta) \cdot \pi^2 \cdot E}{3 \cdot \lambda^2 \cdot \sigma_c} - \sqrt{\left(\frac{1}{2} + \frac{(1+\eta) \cdot \pi^2 \cdot E}{3 \cdot \lambda^2 \cdot \sigma_c} \right)^2 - \frac{\pi^2 \cdot E}{1.5 \cdot \lambda^2 \cdot \sigma_c}}$$

$$K_{12.med} = 0.26$$

(b) Using Table 19:

For $\quad \dfrac{E}{\sigma_{c.g.\text{/\!/}} \cdot K_{2.comp} \cdot K_3} = 924.44 \quad$ and $\quad \lambda = 132.67 \quad$ thus, $K_{12.med} = 0.26$

(2) Very-short term loading (3 s gust):
(a) Using equation in Annex B:

$$K_3 := K_{3.v.short} \qquad \sigma_c := \sigma_{c.g.//} \cdot K_{2.comp} \cdot K_3$$

$$\lambda := \lambda_{yy} \qquad \eta := 0.005 \cdot \lambda \qquad E := E_{min} \cdot K_{2.E}$$

$$K_{12.v.short} := \frac{1}{2} + \frac{(1+\eta) \cdot \pi^2 \cdot E}{3 \cdot \lambda^2 \cdot \sigma_c} - \sqrt{\left(\frac{1}{2} + \frac{(1+\eta) \cdot \pi^2 \cdot E}{3 \cdot \lambda^2 \cdot \sigma_c}\right)^2 - \frac{\pi^2 \cdot E}{1.5 \cdot \lambda^2 \cdot \sigma_c}}$$

$$K_{12.v.short} = 0.2$$

(b) Using Table 19:

For $\dfrac{E}{\sigma_{c.g.//} \cdot K_{2.comp} \cdot K_3} = 660.32$ and $\lambda = 132.67$ thus, $K_{12.v.short} = 0.2$

5. Loading

Uniformly distributed axial load	$P := 5.4 \cdot kN \cdot m^{-1}$
Stud spacing, S_{stud}	$S_{stud} := 0.4 \cdot m$
Axial load per stud, P_{stud}	$P_{stud} := P \cdot S_{stud}$
	$P_{stud} = 2.16 \circ kN$
Applied wind load, WL	$WL := 0.975 \cdot kN \cdot m^{-2}$
Wind load per stud, W_{stud}	$W_{stud} := WL \cdot S_{stud}$
	$W_{stud} = 0.39 \circ kN \cdot m^{-1}$

6. Compression and bending stresses

(1) Medium-term loading (no wind):

Applied compressive stress per stud	$\sigma_{c.a.//} := \dfrac{P_{stud}}{A}$
	$\sigma_{c.a.//} = 0.46 \circ N \cdot mm^{-2}$
Permissible compressive stress	$\sigma_{c.adm.//} := \sigma_{c.g.//} \cdot K_{2.comp} \cdot K_{3.med} \cdot K_8 \cdot K_{12.med}$
	$\sigma_{c.adm.//} = 1.63 \circ N \cdot mm^{-2}$
	Compression satisfactory

(2) Very-short term loading (3 s gust):

Applied compressive stress	As above
	$\sigma_{c.a.//} = 0.46 \circ N \cdot mm^{-2}$
Permissible compressive stress	$\sigma_{c.adm.//} := \sigma_{c.g.//} \cdot K_{2.comp} \cdot K_{3.v.short} \cdot K_8 \cdot K_{12.v.short}$
	$\sigma_{c.adm.//} = 1.77 \circ N \cdot mm^{-2}$
	Compression satisfactory
Applied bending moment per stud	$M := \dfrac{W_{stud} \cdot L_x^2}{8}$
	$M = 0.63 \circ kN \cdot m$

Applied bending stress $\sigma_{m.a.\//} := \dfrac{M}{Z_{xx}}$

$\sigma_{m.a.\//} = 8.07 \circ \mathrm{N \cdot mm^{-2}}$

Permissible bending stress $\sigma_{m.adm.\//} := \sigma_{m.g.\//} \cdot K_{2.bending} \cdot K_{3.v.short} \cdot K_6 \cdot K_7 \cdot K_8$

$\sigma_{m.adm.\//} = 11.82 \circ \mathrm{N \cdot mm^{-2}}$

Check interaction quantity

Euler critical stress $\sigma_e := \dfrac{\pi^2 \cdot E}{\lambda^2}$

$\sigma_e = 2.92 \circ \mathrm{N \cdot mm^{-2}}$

$$\dfrac{\sigma_{m.a.\//}}{\sigma_{m.adm.\//} \cdot \left(1 - \dfrac{1.5 \cdot \sigma_{c.a.\//}}{\sigma_e} \cdot K_{12.v.short}\right)} + \dfrac{\sigma_{c.a.\//}}{\sigma_{c.adm.\//}}$$

$$= 0.98 < 1$$

Interaction quantity is satisfactory

Therefore 47 mm × 100 mm timber sections in strength class C22 are satisfactory

Example 5.5   Axial load capacity of a tensile member with lateral load

A trussed rafter tie of 38 mm × 100 mm section is 2.7 m long and is subjected to a lateral concentrated load of 0.65 kN at mid-length. Determine the maximum medium-term axial tensile load that the rafter tie can carry. Assume a timber in strength class C22 under service class 2 conditions.

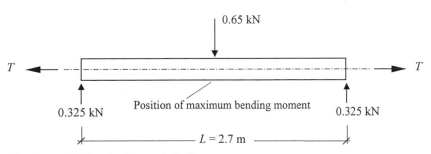

Fig. 5.10   Trussed rafter tie (Example 5.5).

Definitions

Force, kN	$\mathrm{N} := \mathrm{newton}$
Length, m	$\mathrm{kN} := 10^3 \cdot \mathrm{N}$
Cross-sectional dimensions, mm	Direction parallel to grain, //
Stress, Nmm^{-2}	Direction perpendicular to grain, *pp*

1. *Geometrical properties*

BS 5268 : Part 2, Table 18

Tie length, L $L := 2.7 \cdot m$

Effective length, L_e $L_e := 1.0 \cdot L$

$L_e = 2.7 \circ m$

Width of section, b $b := 38 \cdot mm$

Depth of section, h $h := 100 \cdot mm$

Cross-sectional area, A $A := b \cdot h$

$A = 3800 \circ mm^2$

For a rectangular section, $i_{yy} := \dfrac{b}{\sqrt{12}}$

    radius of gyration, i_{yy}

$i_{yy} = 10.97 \circ mm$

Section modulus, Z_{xx} $Z_{xx} := \dfrac{b \cdot h^2}{6}$

$Z_{xx} = 63\,333.33 \circ mm^3$

2. *Check slenderness ratio, λ*

$$\lambda := \frac{L_e}{i_{yy}}$$

$\lambda = 246.13$

< 250, satisfactory

3. *Grade stresses*

BS 5268 : Part 2, Table 7

Timber strength class, C22

Bending parellel to grain $\sigma_{m.g.//} := 6.8 \cdot N \cdot mm^{-2}$

Tension parallel to grain $\sigma_{t.g.//} := 4.1 \cdot N \cdot mm^{-2}$

Minimum modulus of elasticity $E_{min} := 6500 \cdot N \cdot mm^{-2}$

4. *K-factors*

Service class 2 (K_2, Table 13) $K_2 := 1$

Medium-term loading $K_3 := 1.25$
  (K_3, Table 14)

Form factor (K_6, Clause 2.10.5) $K_6 := 1$

Depth factor (K_7, Clause 2.10.6) $K_7 := \left(\dfrac{300 \cdot mm}{h} \right)^{0.11}$

$K_7 = 1.13$

Width factor (K_{14}, Clause 2.12.2) $K_{14} := \left(\dfrac{300 \cdot mm}{h} \right)^{0.11}$

$K_{14} = 1.13$

No load sharing (K_8, Clause 2.9) $K_8 := 1$

5. *Loading*

Lateral point load $P := 0.65 \cdot \text{kN}$

Applied moment $M := \dfrac{P \cdot L_e}{4}$

$M = 0.44 \circ \text{kN} \cdot \text{m}$

6. *Applied and permissible bending stresses*

Applied bending stress $\sigma_{m.a.\parallel} := \dfrac{M}{Z_{xx}}$

$\sigma_{m.a.\parallel} = 6.93 \circ \text{N} \cdot \text{mm}^{-2}$

Permissible bending stress $\sigma_{m\,adm\,\parallel} := \sigma_{m.g.\parallel} \cdot K_2 \cdot K_3 \cdot K_6 \cdot K_7 \cdot K_8$

$\sigma_{m.adm.\parallel} = 9.59 \circ \text{N} \cdot \text{mm}^{-2}$

Bending satisfactory

7. *Permissible tensile stress*

$\sigma_{t.adm.\parallel} := \sigma_{t.g.\parallel} \cdot K_2 \cdot K_3 \cdot K_8 \cdot K_{14}$

$\sigma_{t.adm.\parallel} = 5.78 \circ \text{N} \cdot \text{mm}^{-2}$

8. *Using interaction quantity*

where $\dfrac{\sigma_{m.a.\parallel}}{\sigma_{m.adm.\parallel}} + \dfrac{\sigma_{c.a.\parallel}}{\sigma_{c.adm.\parallel}} \leq 1$

hence $\sigma_{t.a.\parallel} := \sigma_{t.adm.\parallel} \cdot \left(1 - \dfrac{\sigma_{m.a.\parallel}}{\sigma_{m.adm.\parallel}}\right)$

$\sigma_{t.a.\parallel} = 1.61 \circ \text{N} \cdot \text{mm}^{-2}$

Hence the axial tensile load capacity of the tie is:

$T_{capacity} := \sigma_{t.a.\parallel} \cdot A$

$T_{capacity} = 61 \circ \text{kN}$

Chapter 6
Design of Glued Laminated Members

6.1 Introduction

Glued laminated timber, commonly known as *glulam*, is a structural material obtained by gluing together at least four laminations of timber sections of usually 19–50 mm basic thickness so that the grain directions of all are parallel. Normally the laminations are dried to around 12–18% moisture content before being machined and assembled. This leads to a reduction in laminations' thickness of around 3–5 mm. Assembly is commonly carried out by applying carefully controlled adhesive mix to the faces of the laminations. They are then placed in mechanical or hydraulic jigs of the appropriate shape and size, and pressure is applied at right angles to the glue lines and is held until curing of the adhesive is complete. Glulam is then cut, shaped and any specified preservative and finishing treatments are applied. In Fig. 6.1 the cross-section of a typical glulam beam is shown.

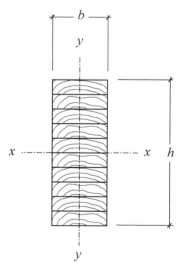

Fig. 6.1 Cross-section of a glulam beam.

Since the length of glulam members normally exceeds the length of commercially available solid timber, the individual laminations are finger-jointed together to make laminations of the required length. These finger-joints are then placed randomly throughout the glulam component.

Timber sections with a thickness of around 33 mm to a maximum of 50 mm are used to laminate *straight* or *slightly curved* members, whereas much thinner sections (12 mm or 19 mm, up to about 33 mm) are used to laminate *curved* members. Glued laminated members can also be constructed with variable sections to produce tapering beams, columns, arches and portals.

A glulam beam designed to resist loads applied perpendicular to the plane of the laminations is referred to as *horizontally laminated*, and when it is designed to resist loads applied parallel to the plane of the laminations it is *vertically laminated*, see Fig. 6.2.

Construction with glulam members offers the advantages of excellent strength and stiffness-to-weight ratio, which are usual with timber materials, and for this reason it has been selected for a number of prestige projects in the last few decades. Glued laminated members can also be designed to have substantial fire resistance. A summary of the advantages of glulam members over solid timber sections is as follows:

(1)   Glulam sections are built up from thin members and it is therefore possible to manufacture complicated shapes. They can be produced in any size, length, and shape, and manufacturing and transportation facilities remain as the only practical limiting factors affecting dimensions.

(2)   The use of a number of laminates enables an even distribution of defects and the location of better quality timber where it is most efficiently utilised.

(3)   It is easily possible to build-in a camber at the time of assembly.

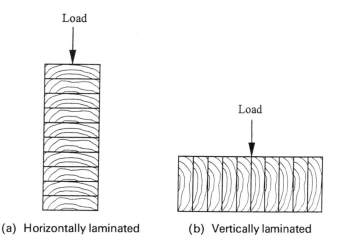

(a) Horizontally laminated   (b) Vertically laminated

Fig. 6.2   Horizontally and vertically laminated glulam beams.

This chapter details the general considerations necessary for the design of flexural and axially loaded glued laminated members.

6.2 Design considerations

The main design considerations for flexural and axially loaded glulam members are the same as those described in detail in Chapters 4 and 5 respectively. It is therefore assumed that the reader is familiar with the various modification factors and all other aspects of the design methodology detailed in these previous chapters. They are not repeated at length here, but when a factor is used, reference is made to the explanatory item of the relevant chapter.

BS 5268 : Part 2 : 1996, Section 3 deals with the design of glued laminated timber and specifies that such timber should be manufactured in accordance with Clause 3.4 of the code and also with the recommendations of BS EN 386 : 1995 *Glued laminated timber. Performance requirements and minimum production requirements*. It also requires that all timber used for laminated softwood members (similar to solid section timber, see Chapter 2, Section 2.1) should be strength graded visually or mechanically in accordance with BS EN 518 or BS EN 519, respectively.

Glulam strength properties may be determined from the strength class properties given in Table 7 of the code (reproduced here as Table 2.3). These are described in detail in the following sections.

6.3 Grade stresses for horizontally glued laminated members

6.3.1 *Single-grade members*

If a glulam section is manufactured using the same grade throughout it is referred to as *single-grade* glued laminated member. The grade stresses given in Tables 7–12 of BS 5268 : Part 2 apply primarily to solid structural timber. In order to determine the grade stresses for a horizontally glued laminated softwood or hardwood member, Clause 3.2 of the code recommends that the strength class stresses given in its Tables 7–12 for the relevant grade and species are multiplied by the modification factors K_{15} to K_{20} according to the nature of stress and the number of laminations. The modification factors K_{15} to K_{20}, detailed in Table 21 of the code, are for *single-grade horizontally glued laminated members* (Table 21 is reproduced here as Table 6.1).

6.3.2 *Combined-grade members*

For economic reasons there are many instances where members may be horizontally laminated with two different strength class timber sections.

Table 6.1 Modification factors K_{15}, K_{16}, K_{17}, K_{18}, K_{19} and K_{20} for single-grade glued laminated members and horizontally glued laminated beams (Table 21, BS 5268)

Strength classes	Number of laminations[a]	Bending // to grain $(\sigma_{m,g,\parallel})$ K_{15}	Tension // to grain $(\sigma_{t,g,\parallel})$ K_{16}	Compression // to grain $(\sigma_{c,g,\parallel})$ K_{17}	Compression $\perp$ to grain[b] $(\sigma_{c,g,\perp})$ K_{18}	Shear // to grain $(\tau_{g,\parallel})$ K_{19}	Modulus of elasticity[c] E_{mean} K_{20}
C27, C30, D50, D60, D70	4 or more	1.39	1.39	1.11	1.49	1.49	1.03
C22, C24, D35, D40	4	1.26	1.26	1.04	1.55	2.34	1.07
	5	1.34	1.34				
	7	1.39	1.39				
	10	1.43	1.43				
	15	1.48	1.48				
	20 or more	1.52	1.52				
C16, C18, D30	4	1.05	1.05	1.07	1.69	2.73	1.17
	5	1.16	1.16				
	7	1.29	1.29				
	10	1.39	1.39				
	15	1.49	1.49				
	20 or more	1.57	1.57				

[a] Interpolation is permitted for intermediate numbers of laminations
[b] K_{18} should be applied to the lower value given in Table 7 for compression perpendicular to grain
[c] K_{20} should be applied to the mean value of modulus of elasticity

Higher-grade laminations are normally placed in the outer portions of the member where they are effectively used, and lower-grade laminations are placed in the inner portion (core) where the lower strength will not greatly affect the overall strength. The combination is permitted provided that the strength classes are not more than three classes apart (as given in Table 7 of the code.) For example C24 and C16 can be horizontally laminated but C24 and C14 should not. The code also requires that in horizontally laminated members where combination grades are used, they should be fabricated so that not less than 25% of the depth at both the top and bottom of the member is of the higher strength class, see Fig. 6.3. For such members the grade stresses should be taken as the product of the strength class stresses (given in Table 7) for the *higher strength class laminations* and the modification factors from Table 21 of the code. In addition, for bending, tension and compression parallel to grain, the grade stresses should be multiplied by 0.95.

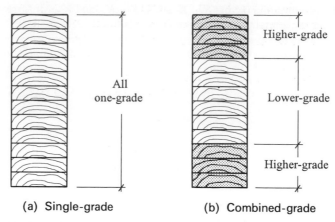

(a) Single-grade   (b) Combined-grade

Fig. 6.3 Horizontally glued laminated members.

6.3.3 *Permissible stresses for horizontally glued laminated members*

The *permissible stresses* for a horizontally glued laminated member are determined as the product of grade stresses, given in Table 7, by any relevant modification factors, i.e. K_2, K_3, ..., K_{14} and modification factors K_{15} to K_{20} appropriate to the number of laminations. These are summarised below:

Bending parallel to grain:
    for a single-grade glulam, $\sigma_{m,adm,\parallel} = \sigma_{m,g,\parallel} \times K_2 K_3 K_6 K_7 K_8 K_{15}$
    for a combined-grade glulam, $\sigma_{m,adm,\parallel} = \sigma_{m,g,\parallel} \times K_2 K_3 K_6 K_7 K_8 K_{15} \times 0.95$

Tension parallel to grain:
    for a single-grade glulam, $\sigma_{t,adm,\parallel} = \sigma_{t,g,\parallel} \times K_2 K_3 K_8 K_{14} K_{16}$
    for a combined-grade glulam, $\sigma_{t,adm,\parallel} = \sigma_{t,g,\parallel} \times K_2 K_3 K_8 K_{14} K_{16} \times 0.95$

Compression parallel to grain:
    for a single-grade glulam, $\sigma_{c,adm,\parallel} = \sigma_{c,g,\parallel} \times K_2 K_3 K_8 K_{12} K_{17}$
    for a combined-grade glulam, $\sigma_{c,adm,\parallel} = \sigma_{c,g,\parallel} \times K_2 K_3 K_8 K_{12} K_{17} \times 0.95$

Compression perpendicular to grain:
    for both single- and combined-grade glulam members,

$$\sigma_{c,adm,\perp} = \sigma_{c,g,\perp} \times K_2 K_3 K_4 K_8 K_{18}$$

Shear parallel to grain:
    for both single- and combined-grade glulam members,

$$\tau_{adm,\parallel} = \tau_{g,\parallel} \times K_2 K_3 K_5 K_8 K_{19}$$

Modulus of elasticity:
    for both single- and combined-grade glulam members,

$$E_{glulam} = E_{mean} \times K_{20}$$

It should also be noted that for tension perpendicular to grain and torsional shear, permissible stresses should be calculated in accordance with Clause 2.7 of the code, and factors K_{19} and K_{27} should be disregarded. For example,

For tension perpendicular to grain:
for both single- and combined-grade glulam members,

$$\sigma_{t,adm,\perp} = \tfrac{1}{3}\tau_{g,/\!/} \times K_2 K_3 K_8$$

In determining the permissible stresses for glued laminated members, the modification factors used (if relevant) are:

K_2 service class 3 (wet exposure condition), see Section 2.5.2,

K_3 duration of loading, see Section 2.5.4,

K_4 bearing stress, see Section 4.5.1,

K_5 shear at notched end, see Section 4.6.1,

K_6 form factor, see Section 4.3.2,

K_7 depth factor, see Section 4.3.3, which applies to the full glulam section size, not the individual laminates,

K_8 load sharing system, see Section 2.5.6,

K_{12} factor for compression member, see Section 5.2.3 (for calculation of K_{12} for glulam compression members, the value of E_{glulum} as defined, should be used, i.e. $E_{glulum} = E_{mean} \times K_{20}$),

K_{14} width factor for tension member, see Section 5.3.2, which applies to the full glulam section size, not the individual laminates,

$K_{15}–K_{20}$ are the appropriate grade stress modification factors for single-grade horizontally laminated members. For combined-grade members, these factors apply to the higher-grade lamination stresses.

Attention is drawn to the definitions for K_7, K_{12} and K_{14} given above.

6.4 Grade stresses for vertically glued laminated beams

A glulam beam designed to resist loads applied parallel to the plane of laminations is referred to as a *vertically laminated beam*. This is illustrated in Fig. 6.2.

BS 5268 : Part 2, Clause 3.3 specifies that permissible stresses for vertically glued laminated beams are governed by the particular conditions of service and loading defined in its Clauses 2.6.2, 2.8 and 2.9 which relate to service class 3 (wet-exposure) condition, duration of loading and load-sharing systems respectively; and also by the additional modification factors K_{27}, K_{28} and K_{29} given in Table 22 of the code appropriate to the number of laminations. Table 22 of the code is reproduced here as Table 6.2.

Table 6.2 Modification factors K_{27}, K_{28}, and K_{29}, for vertically glued laminated members, (Table 22, BS 5268 : Part 2)

Number of laminations	Bending // to grain, ($\sigma_{m,g,//}$), tension // to grain, ($\sigma_{t,g,//}$), and shear // to grain, ($\tau_{g,//}$) K_{27}		Modulus of elasticity, (E_{min})[a] and compression // to grain, ($\sigma_{c,g,//}$) K_{28}		Compresison $\perp$ to grain[b] ($\sigma_{c,g,\perp}$) K_{29}
	Softwoods	Hardwoods	Softwoods	Hardwoods	Softwoods and hardwoods
2	1.11	1.06	1.14	1.06	
3	1.16	1.08	1.21	1.08	
4	1.19	1.10	1.24	1.10	
5	1.21	1.11	1.27	1.11	1.10
6	1.23	1.12	1.29	1.12	
7	1.24	1.12	1.30	1.12	
8 or more	1.25	1.13	1.32	1.13	

[a] When applied to the value of the modulus of elasticity, E, K_{28} is applicable to the minimum value of E

[b] If no wane is present, K_{29} should have the value 1.33, and, regardless of the grade of timber used, should be applied to the SS or HS grade stress for the species

In general it is not recommended that vertically laminated beams be designed using different grade laminates (i.e. combined strength classes). When a designer is considering a glulam beam with vertical laminations, it is often because a horizontally laminated member with bending about the x–x axis is also being subjected to bending on its y–y axis, for example, due to lateral loading. If the vertically laminated beam is made from different grade laminations it can be assumed that such a member has grade stresses equal to the product of *weighted* strength class stresses, and the modification factors K_{27}, K_{28} and K_{29}, given in Table 22, appropriate to the total number of laminations in the beam.

A *weighted* strength class stress, σ_w, may be obtained from:

$$\sigma_w = \sigma_{g,lower} \times \frac{N_{lower}}{N_{total}} + \sigma_{g,higher} \times \frac{N_{higher}}{N_{total}} \tag{6.1}$$

where $\sigma_{g,lower}$ and $\sigma_{g,higher}$ refer to appropriate stress or modulus of elasticity for the lower-grade and higher-grade laminations respectively, as given in Table 7 of the code; and similarly N refers to the appropriate number of laminations in the member.

For example, the weighted bending stress parallel to grain for a vertically laminated beam of 12 laminations, where 3 laminations at each end are in

C24 and 6 inner ones are in C18 softwood timber, may be calculated as follows:

From Table 7 of the code (Table 2.3 in this book):

for C18 timber, $\sigma_{m,g,/\!/} = 5.8\,\text{N/mm}^2$
for C24 timber, $\sigma_{m,g,/\!/} = 7.5\,\text{N/mm}^2$

Therefore the weighted grade bending stress is:

$$\sigma_{w,m,g,/\!/} = 5.8 \times \tfrac{6}{12} + 7.5 \times \tfrac{6}{12} = 6.65\,\text{N/mm}^2$$

Hence the weighted bending stress parallel to grain for this glulam beam is calculated as:

From Table 22 of the code (Table 6.2 in this book), for 12 laminations, $K_{27} = 1.25$.

$$\sigma_{w,m,/\!/} = 6.65 \times 1.25 = 8.31\,\text{N/mm}^2$$

6.4.1 *Permissible stresses for vertically glued laminated members*

The permissible stresses for a vertically glued laminated member are determined as the product of grade stresses (or weighted grade stresses in the case of a member with combined-grade laminations) by any relevant modification factors, i.e. $K_2, K_3, \ldots, K_{14}$ and modification factors K_{27}, K_{28} and K_{29} appropriate to the number of laminations). These are summarised below:

Bending parallel to grain: $\quad \sigma_{m,adm,/\!/} = \sigma_{m,g,/\!/} \times K_2 K_3 K_6 K_7 K_8 K_{27}$
Tension parallel to grain: $\quad \sigma_{t,adm,/\!/} = \sigma_{t,g,/\!/} \times K_2 K_3 K_8 K_{14} K_{27}$
Shear parallel to grain: $\quad \tau_{adm,/\!/} = \tau_{g,/\!/} \times K_2 K_3 K_5 K_8 K_{27}$
Compression parallel to grain: $\quad \sigma_{t,adm,/\!/} = \sigma_{t,g,/\!/} \times K_2 K_3 K_8 K_{12} K_{28}$
Compression perpendicular to grain: $\quad \sigma_{c,adm,\perp} = \sigma_{c,g,\perp} \times K_2 K_3 K_4 K_8 K_{29}$
Modulus of elasticity: $\quad E_{glulam} = E_{mean}$

For vertically laminated beams, the E value (for 4 or more laminations) is E_{mean} and K_{28} is only applicable to E_{min}. It should be noted that the size factors K_7 and K_{14} for all glulam members, whether horizontally or vertically laminated, should be applied to glulam strengths not to the individual laminate strengths.

It should also be noted that for tension perpendicular to grain and torsional shear, permissible stresses should be calculated in accordance with Clause 2.7 of the code, and factor K_{27} should be disregarded.

Lateral torsional buckling is unlikely to be a problem as the depth-to-breadth ratio related to bending about the y–y axis will almost certainly be less than 1.0 and any beam tends to buckle only at right angles to the direction of bending.

6.5   Deformation criteria for glued laminated beams

BS 5268 : Part 2 : 1996, Clause 3.5 recommends that in addition to the deflection due to bending, the shear deflection may be significant and should be taken into account. Attention is drawn to deflection criteria for flexural members which were covered in detail in Section 4.4, as the same criteria apply to glulam beams.

As it is easily possible to build-in a *camber* at the time of assembly, the code specifies that members can be *pre-cambered* to offset the deflection under dead or permanent loads. In such cases the deflection under live or intermittent imposed load should not exceed 0.003 of the span.

6.6   Curved glued laminated beams

Glued laminated beams are often curved and/or tapered in order to meet architectural requirements, to provide pitched roofs, to obtain maximum interior clearance and/or to reduce wall height requirements at the end supports. The most commonly used types are the curved beams with constant rectangular cross-section and the double tapered beams (see Fig. 6.4).

The distribution of bending stress, σ_m, in a curved beam is non-linear and, in addition, radial stresses, σ_r, perpendicular to the grain are caused by bending moments. If the bending moment tends to increase the radius of curvature, the radial stresses are in tension perpendicular to the grain (i.e. $\sigma_r = \sigma_{t,\perp}$); and if it tends to decrease the radius of curvature, the radial stresses are in compression perpendicular to the grain (i.e. $\sigma_r = \sigma_{c,\perp}$).

Curved glulam beams are formed by bending laminations in purpose-built jigs during the production process. In order to make allowance for the initial stresses induced in the laminations, a modification factor K_{33} has

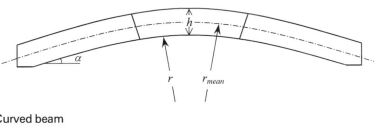

(a) Curved beam

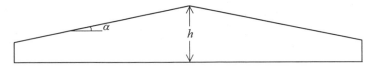

(b) Double tapered beam

Fig. 6.4   Curved and double tapered beams.

been introduced. For curved glulam beams with constant cross-section, BS 5268 : Part 2, Clause 3.5.3 specifies the following requirements:

(1) The ratio of the radius of curvature, r, to the lamination thickness, t, should be greater than the mean modulus of elasticity, in N/mm^2, for the strength class divided by 70, for both softwoods and hardwoods (see Fig. 6.5). Therefore check that:

$$\frac{r}{t} > \frac{E_{mean}}{70}$$

(2) If $\frac{r}{t} < 240$, then the bending, tension and compression grade stresses parallel to grain should be multiplied by the modification factor K_{33}, where

$$K_{33} = 0.75 + 0.001\,\frac{r}{t} \quad \text{and} \quad K_{33} \leq 1.0$$

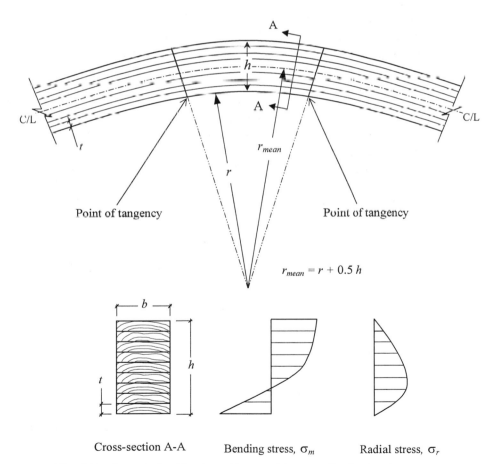

Cross-section A-A   Bending stress, σ_m Radial stress, σ_r

Fig. 6.5 Curved glued laminated beam with stress distributions at apex.

(3)   In curved beams, where the ratio of the minimum mean radius of curvature, r_{mean}, to the depth, h, is less than or equal to 15 (i.e. if $\dfrac{r_{mean}}{h} \leq 15$), then the bending stress induced by a moment, M, should be calculated as:

(a)   Bending stress in the extreme fibre on the concave face,

$$\sigma_m = k_{34} \frac{6M}{bh^2} \tag{6.2}$$

where:

$$K_{34} = 1 + \left(0.5 \frac{h}{r_{mean}}\right) \quad \text{for } \frac{r_{mean}}{h} \leq 10, \text{ and}$$

$$K_{34} = 1.15 - \left(1.01 \frac{r_{mean}}{h}\right) \quad \text{for } 10 < \frac{r_{mean}}{h} \leq 15$$

(b)   Bending stress in the extreme fibre on the convex face,

$$\sigma_m = \frac{6M}{bh^2} \tag{6.3}$$

(4)   The radial stress, σ_r, induced by a bending moment, M, should be calculated as:

$$\sigma_r = \frac{3M}{2bhr_{mean}} \tag{6.4}$$

where:

(a)   If the moment tends to increase the radius of curvature of the beam, the radial stress will be tension perpendicular to the grain and the value of σ_r should not be greater than the value derived in accordance with Clauses 3.2 and 2.7 of the code. Therefore:

$$\sigma_r \leq \sigma_{t,adm,\perp} = \tfrac{1}{3}\tau_{g,/\!/} \times K_2 K_3 K_8 \tag{6.5}$$

where, from Section 2.5.7, $\sigma_{t,g,\perp} = \tfrac{1}{3}\tau_{g,/\!/}$.

(b)   If the moment tends to reduce the radius of curvature of the beam, the radial stress will be compression perpendicular to the grain and the value of σ_r should not be greater than 1.33 times the compression perpendicular to the grain stress for the strength class. Therefore:

$$\sigma_r \leq 1.33 \times \sigma_{c,adm,\perp} = 1.33 \times \sigma_{c,g,\perp} \times K_2 K_3 K_4 K_8 K_{18} \tag{6.6}$$

6.7   Bibliography

Baird and Ozelton (1984) *Timber Designer's Manual*, 2nd ed. Blackwell Science, Oxford.

Mettem, C.J. (1986) *Structural Timber Design and Technology*. Longman Scientific and Technical, Harlow.

TRADA (1995) *Glued laminated timber – an introduction*, *Wood Information*, *Section 1*, *Sheet 6*. Timber Research and Development Association, High Wycombe, UK.

6.8 Design examples

Example 6.1 Design of a single-grade glued laminated beam

A series of glulam beams of 115 mm wide × 560 mm deep (14 laminations) with an effective span of 10.5 m is to be used in the construction of the roof of an exhibition hall. The roof comprises exterior tongued and grooved solid softwood decking exposed on the underside and covered on the top with 3 layer felt and chippings. For the design data given below,

(a) check the suitability of the section,
(b) determine the camber required.

Design data

Timber	Imported whitewood in strength class C24 and service class 1
Bearing width	90 mm each end
Dead load (long-term):	
Exterior decking	0.75 kN/m
3 layer felt and chippings	0.63 kN/m
Beam self-weight	0.3 kN/m
Imposed load (medium-term)	2.5 kN/m

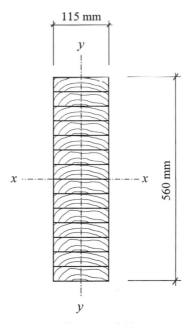

Fig. 6.6 Cross-section of glulam beam (Example 6.1).

Definitions

Force, kN	$N :=$ newton
Length, m	$kN := 10^3 \cdot N$
Cross-sectional dimensions, mm	Direction parallel to grain, $//$
Stress, Nmm^{-2}	Direction perpendicular to grain, pp

1. Geometrical properties

Effective span, L_e	$L_e := 10.5 \cdot m$
Bearing width, bw	$bw := 90 \cdot mm$
Beam dimensions:	
Breadth of section, b	$b := 115 \cdot mm$
Depth of section, h	$h := 560 \cdot mm$
Cross-sectional area, A	$A := b \cdot h$
	$A = 6.44 \times 10^4 \circ mm^2$
Second moment of area, I_{xx}	$I_{xx} := \frac{1}{12} \cdot b \cdot h^3$
	$I_{xx} = 1.68 \times 10^9 \circ mm^4$

2. Loading

Dead load:	
Self-weight of beam (kN/m), Swt	$Swt := 0.3 \cdot kN \cdot m^{-1}$
t and g decking, (kN/m), tg	$tg := 0.75 \cdot kN \cdot m^{-1}$
3 layer felt and chippings (kN/m), $fchip$	$fchip := 0.63 \cdot kN \cdot m^{-1}$
Total dead load, DL	$DL := (Swt + tg + fchip) \cdot L_e$
	$DL = 17.64 \circ kN$
Imposed load:	
Prescribed imposed load, (kN/m), pil	$pil := 2.5 \cdot kN \cdot m^{-1}$
Total imposed load, IL	$IL := pil \cdot L_e$
	$IL = 26.25 \circ kN$
Total load (kN):	
Long-term, W_{long}	$W_{long} := DL$
	$W_{long} = 17.64 \circ kN$
Medium-term, W_{med}	$W_{med} := DL + IL$
	$W_{med} = 43.89 \circ kN$

3. K-factors

Service class 1 (K_2, Table 13)	$K_2 := 1$
Medium-term loading (K_3, Table 14)	$K_3 := 1.25$
Bearing: 90 mm (K_4, Table 15) (bearing position not defined)	$K_4 := 1.0$

Notched end effect
(K_5, Clause 2.10.4) $K_5 := 1$

Form factor (K_6, Clause 2.10.5) $K_6 := 1$

Depth factor (K_7, Clause 2.10.6)
to provide a dimensionless
value for K_7 in Mathcad, let: $h_1 := h \cdot mm^{-1}$

for $h > 300 \, mm$ $K_7 := 0.81 \cdot \dfrac{h_1^2 + 92\,300}{h_1^2 + 56\,800}$

$K_7 = 0.89$

No load sharing (K_8, Clause 2.9) $K_8 := 1$

Glulam modification factors (Table 21):
Single-grade, 15 laminates in
C24 timber

Bending // to grain, K_{15} $K_{15} := 1.48$

Compression perpendicular $K_{18} := 1.55$
to grain, K_{18}

Shear // to grain, K_{19} $K_{19} := 2.34$

Modulus of elasticity, K_{20} $K_{20} := 1.07$

4. Grade stresses

BS 5268 : Part 2, Table 7
Timber strength class, C24

Bending // to grain $\sigma_{m.g.\!/\!/} := 7.5 \cdot N \cdot mm^{-2}$

Compression perpendicular $\sigma_{c.g.pp} := 1.9 \cdot N \cdot mm^{-2}$
to grain

Shear // to grain $\tau_{g.\!/\!/} := 0.71 \cdot N \cdot mm^{-2}$

Mean modulus of elasticity $E_{mean} := 10\,800 \cdot N \cdot mm^{-2}$

5. Bending stress

Applied bending moment $M_{med} := \dfrac{W_{med} \cdot L_e}{8}$

$M_{med} = 57.61 \circ kN \cdot m$

Section modulus $Z_{xx} := \dfrac{b \cdot h^2}{6}$

$Z_{xx} = 6.01 \times 10^6 \circ mm^3$

Applied bending stress $\sigma_{m.a.\!/\!/} := \dfrac{M_{med}}{Z_{xx}}$

$\sigma_{m.a.\!/\!/} = 9.58 \circ N \cdot mm^{-2}$

Permissible bending stress $\sigma_{m.adm.\!/\!/} := \sigma_{m.g.\!/\!/} \cdot K_2 \cdot K_3 \cdot K_6 \cdot K_7 \cdot K_8 \cdot K_{15}$

$\sigma_{m.adm.\!/\!/} = 12.32 \circ N \cdot mm^{-2}$

Bending stress satisfactory

6. *Lateral stability*

Clause 2.10.8 and Table 16

Maximum depth-to-breadth
  ratio, h/b

$$\frac{h}{b} = 4.87$$

Ends should be held in position and compression edges held in line by direct connection of t & g decking to the beams

7. *Shear stress*

Applied shear force

$$F_v := \frac{W_{med}}{2}$$

$$F_v = 21.95 \circ kN$$

Applied shear stress

$$\tau_{a.\!/\!/} := \frac{3}{2} \cdot \left(\frac{F_v}{b \cdot h}\right)$$

$$\tau_{a.\!/\!/} = 0.51 \circ N \cdot mm^{-2}$$

Permissible shear stress

$$\tau_{adm.\!/\!/} := \tau_{g.\!/\!/} \cdot K_2 \cdot K_3 \cdot K_5 \cdot K_8 \cdot K_{19}$$
$$\tau_{adm.\!/\!/} = 2.08 \circ N \cdot mm^{-2}$$
Shear stress satisfactory

8. *Bearing stress*

Applied load

$$F := \frac{W_{med}}{2}$$

$$F = 21.95 \circ kN$$

Applied bearing stress,
  (bearing area $= b \cdot bw$)

$$\sigma_{c.a.pp} := \frac{F}{b \cdot bw}$$

$$\sigma_{c.a.pp} = 2.12 \circ N \cdot mm^{-2}$$

Permissible bearing stress

$$\sigma_{c.adm.pp} := \sigma_{c.g.pp} \cdot K_2 \cdot K_3 \cdot K_4 \cdot K_8 \cdot K_{18}$$
$$\sigma_{c.adm.pp} = 3.68 \circ N \cdot mm^{-2}$$
Bearing stress satisfactory

9. *Deflection*

Modulus of elasticity of
  glulam, E

$$E := E_{mean} \cdot K_{20}$$
$$E = 1.16 \times 10^4 \circ N \cdot mm^{-2}$$

Deflection due to bending

$$\Delta_m := \frac{5 \cdot W_{med} \cdot L_e^3}{384 \cdot E \cdot I_{xx}}$$

$$\Delta_m = 34.02 \circ mm$$

Deflection due to shear

$$\Delta_s := \frac{19.2 \cdot M_{med}}{(b \cdot h) \cdot E}$$

$$\Delta_s = 1.49 \circ mm$$

Total deflection

$$\Delta_{total} := \Delta_m + \Delta_s$$
$$\Delta_{total} = 35.5 \circ mm$$

Permissible deflection
$$\Delta_{adm} := 0.003 \cdot L_e$$
$$\Delta_{adm} = 31.5 \circ mm$$

Camber required = deflection under permanent (long-term) loading:
$$\Delta_{long} := \frac{5 \cdot W_{long} \cdot L_e^3}{384 \cdot E \cdot I_{xx}}$$
$$\Delta_{long} = 13.67 \circ mm$$
Provide 15 mm camber

Deflection under live load
$$\Delta_{live} := \Delta_{total} - \Delta_{long}$$
$$\Delta_{live} = 21.83 \circ mm$$
Deflection satisfactory

Therefore 115 mm × 560 mm glulam sections in single-grade C24 are satisfactory

Example 6.2   Design of a combined-grade glued laminated beam – selection of a suitable section size

Design of a glued laminated timber beam for the roof of a restaurant is required. The beam is to span 9.8 m centre to centre on 125 mm wide bearings under service class 2 conditions. It is proposed to use a combined-grade lay-up using softwood timber in strength classes of C18 and C16 with laminations of 36 mm finished thickness. The beam is subjected to a dead load of 0.67 kN/m excluding self-weight, from t & g boarding and roofing, and an imposed medium-term load of 2.25 kN/m.

Definitions

Force, kN	N := newton
Length, m	kN := $10^3 \cdot$ N
Cross-sectional dimensions, mm	Direction parallel to grain, //
Stress, Nmm^{-2}	Direction perpendicular to grain, *pp*

1.   Geometrical properties

Effective span, L_e $L_e := 9.8 \cdot$ m
Bearing width, *bw* $bw := 125 \cdot$ mm
Beam dimensions:
Breadth of section, *b*
Depth of section, *h*   } Not known at this stage
Cross-sectional area, *A*
Second moment of area, I_{xx}

2.   Loading

Dead load:
Self-weight of beam $Swt := 0.3 \cdot kN \cdot m^{-1}$ assumed
  (kN/m), *Swt*

Prescribed dead load (kN/m), PDL	$PDL := 0.67 \cdot kN \cdot m^{-1}$
Total dead load, DL	$DL := (Swt + PDL) \cdot L_e$ $DL = 9.51 \circ kN$
Imposed load: Prescribed imposed load (kN/m), pil	$pil := 2.25 \cdot kN \cdot m^{-1}$
Total imposed load, IL	$IL := pil \cdot L_e$ $IL = 22.05 \circ kN$

Total load (kN):

Long-term, W_{long}	$W_{long} := DL$ $W_{long} = 9.51 \circ kN$
Medium-term, W_{med}	$W_{med} := DL + IL$ $W_{med} = 31.56 \circ kN$

3. K-factors

Service class 2 (K_2, Table 13)	$K_2 := 1$
Medium-term loading (K_3, Table 14)	$K_3 := 1.25$
Bearing: 125 mm (K_4, Table 15) (bearing position not defined)	$K_4 := 1.0$
Notched end effect (K_5, Clause 2.10.4)	$K_5 := 1$
Form factor (K_6, Clause 2.10.5)	$K_6 := 1$
Depth factor (K_7, Clause 2.10.6)	Since section size is not known, at this stage assume, $K_7 := 1$
No load sharing (K_8, Clause 2.9)	$K_8 := 1$

Glulam modification factors (Table 21):

BS 5268 : Part 2, Clause 3.2	Stress values for the higher-grade timber apply
Table 7, for C18 timber	Since number of laminations is not known, ignore at this stage
Bending // to grain, K_{15} Compression perpendicular to grain, K_{18} Shear // to grain, K_{19} Modulus of elasticity, K_{20}	$K_{20} := 1.17$

4. Grade stresses

BS 5268 : Part 2, Table 7

Timber strength class, C18	
Bending // to grain	$\sigma_{m.g.//} := 5.8 \cdot N \cdot mm^{-2}$
Compression perpendicular to grain	$\sigma_{c.g.pp} := 2.2 \cdot N \cdot mm^{-2}$

Shear // grain $\qquad$ $\tau_{g.//} := 0.67 \cdot N \cdot mm^{-2}$
Mean modulus of elasticity $\qquad$ $E_{mean} := 9100 \cdot N \cdot mm^{-2}$

5. *Selecting a trial section*

(1) Using deflection criteria:

Modulus of elasticity for $\qquad$ $E := E_{mean} \cdot K_{20}$
glulam, E $\qquad$ $E = 1.06 \times 10^4 \circ N \cdot mm^{-2}$
Permissible deformation $\qquad$ $\Delta_{adm} := 0.003 \cdot L_e$
under imposed load $\qquad$ $\Delta_{adm} = 29.4 \circ mm$

Required second moment of $\qquad$ $I_{xx.rqd} := \dfrac{5 \cdot W_{med} \cdot L_e^3}{384 \cdot E \cdot \Delta_{adm}}$
area, I_{xx} using Δ_{adm}
under imposed load $\qquad$ $I_{xx.rqd} = 1.24 \times 10^9 \circ mm^4$

(2) Using lateral stability criteria:

In order to achieve lateral stability by direct fixing of decking to beams, the depth-to-breadth ratio should be limited to 5, i.e. $h \le 5b$. Substituting for $h = 5b$ in $I_{xx} = bh^3/12$ and equating it to $I_{xx,rqd}$:

$I_{xx.rqd} = \dfrac{b \cdot (5 \cdot b)^3}{12}$ will give $b := \left(\dfrac{12 \cdot I_{xx.rqd}}{5^3} \right)^{1/4}$

$\qquad\qquad\qquad\qquad\qquad\qquad b = 104.36 \circ mm$

and depth of section, h $\qquad$ $h := 5 \cdot b$
$\qquad\qquad\qquad\qquad\qquad\qquad h = 521.79 \circ mm$

Try a 110 mm × 540 mm deep section with 15 laminations of 36 mm in thickness

Trial beam dimensions:
Depth (mm), h $\qquad$ $h := 540 \cdot mm$
Breadth (mm), b $\qquad$ $b := 110 \cdot mm$
Modification factor K_7 for
   section depth:
to provide a dimensionless value
for K_7 in Mathcad, let: $\qquad$ $h_1 := h \cdot mm^{-1}$

for $h > 300$ mm $\qquad$ $K_7 := 0.81 \cdot \dfrac{h_1^2 + 92\,300}{h_1^2 + 56\,800}$

$\qquad\qquad\qquad\qquad\qquad\qquad K_7 = 0.89$

Glulam modification factors (Table 21):
BS 5268 : Part 2, Clause 3.2 $\qquad$ Stress values for the higher-graded timber apply
Combined-grade, 8 laminates in
   C18 timber, by interpolation
Bending // to grain, K_{15} $\qquad$ $K_{15} := 1.32$
Compression perpendicular to $\qquad$ $K_{18} := 1.69$
   grain, K_{18}
Shear // to grain, K_{19} $\qquad$ $K_{19} := 2.73$
Modulus of elasticity, K_{20} $\qquad$ $K_{20} := 1.17$

6. Bending stress

Applied bending moment

$$M_{med} := \frac{W_{med} \cdot L_e}{8}$$

$$M_{med} = 38.66 \circ kN \cdot m$$

Section modulus

$$Z_{xx} := \frac{b \cdot h^2}{6}$$

$$Z_{xx} = 5.35 \times 10^6 \circ mm^3$$

Applied bending stress

$$\sigma_{m.a.\//} := \frac{M_{med}}{Z_{xx}}$$

$$\sigma_{m.a.\//} = 7.23 \circ N \cdot mm^{-2}$$

Permissible bending stress

$$\sigma_{m.adm.\//} := \sigma_{m.g.\//} \cdot K_2 \cdot K_3 \cdot K_6 \cdot K_7 \cdot K_8 \cdot K_{15} \cdot 0.95$$
$$\sigma_{m.adm.\//} = 9.09 \circ N \cdot mm^{-2}$$
Bending stress satisfactory

Check self-weight:
BS 5268 : Part 2, Table 7
Density
Beam self-weight (kN/m):
Actual

$$\rho := 380 \cdot kg \cdot m^{-3}$$

$$Swt_{actual} := \rho \cdot g \cdot h \cdot b$$
$$Swt_{actual} = 0.22 \circ kN \cdot m^{-1}$$

Assumed

$$Swt_{assumed} := Swt$$
$$Swt_{assumed} = 0.3 \circ kN \cdot m^{-1}$$
Self-weight satisfactory

7. Lateral stability

Clause 2.10.8 and Table 16

Maximum depth-to-breadth
  ratio, h/b

$$\frac{h}{b} = 4.91$$

Ends should be held in position and
  compression edges held in line by direct
  connection of t & g decking to beams

8. Shear stress

Applied shear force

$$F_v := \frac{W_{med}}{2}$$

$$F_v = 15.78 \circ kN$$

Applied shear stress

$$\tau_{a.ll} := \frac{3}{2} \cdot \left(\frac{F_v}{b \cdot h} \right)$$

$$\tau_{a.ll} = 0.4 \circ N \cdot mm^{-2}$$

Permissible shear stress

$$\tau_{adm.\//} := \tau_{g.\//} \cdot K_2 \cdot K_3 \cdot K_5 \cdot K_8 \cdot K_{19}$$
$$\tau_{adm.\//} = 2.29 \circ N \cdot mm^{-2}$$
Shear stress satisfactory

9. *Bearing stress*

Applied load

$$F := \frac{W_{med}}{2}$$

$$F = 15.78 \circ \text{kN}$$

Applied bearing stress
(bearing area $= b \cdot bw$)

$$\sigma_{c.a.pp} := \left(\frac{F}{b \cdot bw} \right)$$

$$\sigma_{c.a.pp} = 1.15 \circ \text{N} \cdot \text{mm}^{-2}$$

Permissible bearing stress

$$\sigma_{c.adm.pp} := \sigma_{c.g.pp} \cdot K_2 \cdot K_3 \cdot K_4 \cdot K_8 \cdot K_{18}$$

$$\sigma_{c.adm.pp} = 4.65 \circ \text{N} \cdot \text{mm}^{-2}$$

Bearing stress satisfactory

10. *Deflection*

Modulus of elasticity for
glulam, E

$$E := E_{mean} \cdot K_{20}$$

$$E = 1.06 \times 10^4 \circ \text{N} \cdot \text{mm}^{-2}$$

Second moment of area

$$I_{xx} := \frac{b \cdot h^3}{12}$$

$$I_{xx} = 1.44 \times 10^9 \circ \text{mm}^4$$

Deflection due to bending

$$\Delta_m := \frac{5 \cdot W_{med} \cdot L_e^3}{384 \cdot E \cdot I_{xx}}$$

$$\Delta_m = 25.16 \circ \text{mm}$$

Deflection due to shear

$$\Delta_s := \frac{19.2 \cdot M_{med}}{(b \ h) \cdot E}$$

$$\Delta_s = 1.17 \circ \text{mm}$$

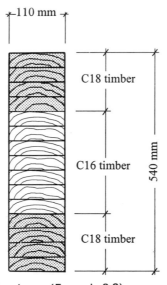

←110 mm→

C18 timber

C16 timber

540 mm

C18 timber

Fig. 6.7 Cross-section of glulam beam (Example 6.2).

Total deflection	$\Delta_{total} := \Delta_m + \Delta_s$
	$\Delta_{total} = 26.34 \circ \text{mm}$
Permissible deflection	$\Delta_{adm} := 0.003 \cdot L_e$
	$\Delta_{adm} = 29.4 \circ \text{mm}$
	Deflection satisfactory and no pre-cambering is required

Therefore adopt a 110 mm × 540 mm glulam section in combined-grades of C18 and C16

Example 6.3   Design of a flooring system

Design of a flooring system using glued laminated timber beams for the first floor of an art gallery is required. The flooring arrangement comprises main beams with effective span of 9.6 m at 4.5 m centres supporting secondary beams at 0.8 m centres and 38 mm tongued and grooved (t & g) flooring using hardwood timber. The secondary beams are simply supported on 100 mm hangers attached to main beams which are in turn supported on load-bearing walls providing a 175 mm bearing width at each end.

It is proposed to use softwood timber in strength class of C24 with laminations of 38 mm finished thickness, single-grade and horizontally glued laminated throughout. The floor is subjected to a dead load of 0.47 kN/m^2, excluding self-weight, and an imposed medium-term load of 3.25 kN/m^2.

Definitions

Force, kN	$N := \text{newton}$
Length, m	$kN := 10^3 \cdot N$
Cross-sectional dimensions, mm	Direction parallel to grain, //
Stress, Nmm^{-2}	Direction perpendicular to grain, *pp*

A.   Design of tongued and grooved boarding

Assuming t & g boarding comprises 100 mm wide timber beams of thickness (depth) 38 mm simply supported on joists.

1.   Loading

Secondary beam spacing, *Js*	$Js := 0.8 \cdot \text{m}$
t & g width, *b*	$b := 100 \cdot \text{mm}$
t & g thickness, *t*	$t := 38 \cdot \text{mm}$
Dead load:	
BS 5268 : Part 2, Table 7	
Average density, D30 hardwood	$\rho_{mean} := 640 \cdot \text{kg} \cdot \text{m}^{-3}$
t & g boarding, self-weight	$tg := \rho_{mean} \cdot g \cdot t$
(kN/m^2), *tg*	$tg = 0.24 \circ \text{kN} \cdot \text{m}^{-2}$

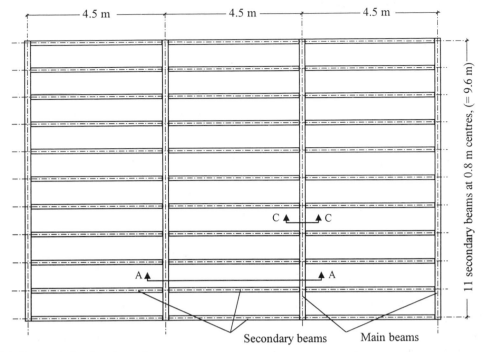

(a) Support structure layout

4.5 m 4.5 m 4.5 m

11 secondary beams at 0.8 m centres, (= 9.6 m)

C▲ ▲C

A▲ ▲A

Secondary beams Main beams

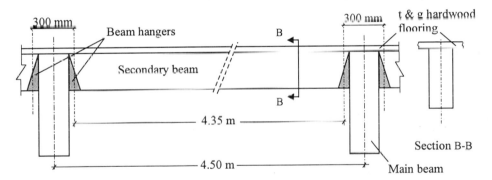

300 mm 300 mm t & g hardwood flooring

Beam hangers B

Secondary beam

B

4.35 m

4.50 m

Section B-B

Main beam

(b) Section A–A: secondary beam details

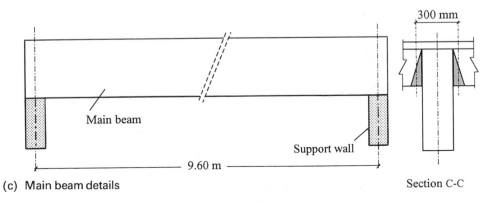

300 mm

Main beam

Support wall

9.60 m

Section C-C

(c) Main beam details

Fig. 6.8 Details of flooring system (Example 6.3).

Prescribed dead load
 (kN/m^2), PDL

$$PDL := 0.47 \cdot kN \cdot m^{-2}$$

Total dead load (kN), DL

$$DL := PDL + tg$$
$$DL = 0.71 \circ kN \cdot m^{-2}$$

Imposed load (kN/m^2), IL

$$IL := 3.25 \cdot kN \cdot m^{-2}$$

Total load (kN), W

$$W := (DL + IL) \cdot b \cdot Js$$
$$W = 0.32 \circ kN$$

2. K-factors

Service class 1 (K_2, Table 13)

$$K_2 := 1$$

Medium-term loading
 (K_3, Table 14)

$$K_3 := 1.25$$

Bearing (K_4, Clause 2.10.2)

$$K_4 := 1.0 \quad \text{assumed}$$

Notched end effect
 (K_5, Clause 2.10.4)

$$K_5 := 1$$

Form factor (K_6, Clause 2.10.5)

$$K_6 := 1$$

Depth factor (K_7, Clause 2.10.6)
 for $h \leq 72$ mm

$$K_7 := 1.17$$

Load sharing applies
 (K_8, Clause 2.9)

$$K_8 := 1.1$$

3. Grade stresses

BS 5268 : Part 2, Table 7
Strength classification
Bending // to grain
Mean modulus of elasticity

Strength class D30
$$\sigma_{m.g.\//} := 9.0 \cdot N \cdot mm^{-2}$$
$$E_{mean} := 9500 \cdot N \cdot mm^{-2}$$

4. Bending stress

Applied bending moment

$$M := \frac{W \cdot Js}{8}$$
$$M = 0.03 \circ kN \cdot m$$

Permissible bending stress

$$\sigma_{m.adm.\//} := \sigma_{m.g.\//} \cdot K_2 \cdot K_3 \cdot K_6 \cdot K_7 \cdot K_8$$
$$\sigma_{m.adm.\//} = 14.48 \circ N \cdot mm^{-2}$$

Required section modulus

$$Z_{rqd} := \frac{M}{\sigma_{m.adm.\//}}$$
$$Z_{rqd} = 2.19 \times 10^3 \circ mm^3$$

Since $Z = \dfrac{b \cdot t^2}{6}$, therefore

$$t_{rqd} := \sqrt{\frac{6 \cdot Z_{rqd}}{b}}$$

$$t_{rqd} = 11.46 \circ mm$$
Thickness satisfactory

5. *Check deflection*

Load sharing system

$E := E_{mean}$

Permissible deflection

$\Delta_{adm} := 0.003 \cdot Js$

$\Delta_{adm} = 2.4 \circ \text{mm}$

Second moment of area

$I_{xx} := \dfrac{b \cdot t^3}{12}$

$I_{xx} = 4.57 \times 10^5 \circ \text{mm}^4$

Deflection

$\Delta := \dfrac{5 \cdot W \cdot Js^3}{384 \cdot E \cdot I_{xx}}$

$\Delta = 0.49 \circ \text{mm}$

Deflection satisfactory

Shear and bearing stresses are not critical in decking arrangements.

Therefore 38 mm t & g boarding in D30 timber is satisfactory

B. *Design of secondary beams*

1. *Loading*

Effective length, L_e

$L_e := 4.35 \cdot \text{m}$

Bearing width, bw

$bw := 100 \cdot \text{mm}$

Beam dimensions:

Depth, h

Breadth, b

$\left.\begin{array}{c} \\ \\ \end{array}\right\}$ Not known at this stage

Total dead load from flooring (kN/m²), DL

$DL = 0.71 \circ \text{kN} \cdot \text{m}^{-2}$

Self-weight of beam (kN/m), Swt

$Swt := 0.3 \cdot \text{kN} \cdot \text{m}^{-1}$ assumed

Imposed load (kN/m²), IL

$IL := 3.25 \cdot \text{kN} \cdot \text{m}^{-2}$

Total load (kN):

Long-term, W_{long}

$W_{long} := \left(DL + \dfrac{Swt}{Js} \right) \cdot L_e \cdot Js$

$W_{long} = 3.77 \circ \text{kN}$

Medium-term, W_{med}

$W_{med} := \left(DL + \dfrac{Swt}{Js} + IL \right) \cdot L_e \cdot Js$

$W_{med} = 15.08 \circ \text{kN}$

2. *K-factors*

Service class 1 (K_2, Table 13)

$K_2 := 1$

Medium-term loading (K_3, Table 14)

$K_3 := 1.25$

Bearing: 100 mm (K_4, Clause 2.10.2)

$K_4 := 1$

Notched end effect $\qquad$ $K_5 := 1$
  (K_5, Clause 2.10.4)
Form factor (K_6, Clause 2.10.5) $\quad$ $K_6 := 1$
Depth factor (K_7, Clause 2.10.6) $\quad$ $K_7 := 1$ $\quad$ assumed
No load sharing (K_8, Clause 2.9) $\quad$ $K_8 := 1$

Glulam modification factors (Table 21):
BS 5268 : Part 2, Clause 3.2
Bending // to grain, K_{15}
Compression perpendicular $\qquad\qquad$ $\left.\begin{array}{l}\\ \\ \\ \\ \end{array}\right\}$ Since number of laminations is not known,
  to grain, K_{18} $\qquad\qquad\qquad\qquad\qquad\qquad$ ignore at this stage
Shear // to grain, K_{19}
Modulus of elasticity, K_{20} $\qquad$ $K_{20} := 1.07$

3. Grade stresses

BS 5268 : Part 2, Clause 3.2
Table 7, for C24 timber
Bending // to grain $\qquad\qquad$ $\sigma_{m.g.\!/\!/} := 7.5 \cdot \mathrm{N} \cdot \mathrm{mm}^{-2}$
Compression perpendicular $\qquad$ $\sigma_{c.g.pp} := 1.9 \cdot \mathrm{N} \cdot \mathrm{mm}^{-2}$
  to grain
Shear // to grain $\qquad\qquad\quad$ $\tau_{g.\!/\!/} := 0.71 \cdot \mathrm{N} \cdot \mathrm{mm}^{-2}$
Mean modulus of elasticity $\qquad$ $E_{mean} := 10\,800 \cdot \mathrm{N} \cdot \mathrm{mm}^{-2}$

4. Selecting a trial section

(1) Using deflection criteria:
   Modulus of elasticity for $\qquad$ $E := E_{mean} \cdot K_{20}$
     glulam, E $\qquad\qquad\qquad\quad$ $E = 1.16 \times 10^4 \circ \mathrm{N} \cdot \mathrm{mm}^{-2}$
   Permissible deformation $\qquad$ $\Delta_{adm} := 0.003 \cdot L_e$
   $\qquad\qquad\qquad\qquad\qquad\quad$ $\Delta_{adm} = 13.05 \circ \mathrm{mm}$

   Required second moment $\qquad$ $I_{xx.rqd} := \dfrac{5 \cdot W_{med} \cdot L_e^3}{384 \cdot E \cdot \Delta_{adm}}$
   of area, I_{xx}, using Δ_{adm}
   $\qquad\qquad\qquad\qquad\qquad\quad$ $I_{xx.rqd} = 1.07 \times 10^8 \circ \mathrm{mm}^4$

(2) Using lateral stability criteria:
   In order to achieve lateral stability by direct fixing of decking to beams, the depth-to-breadth ratio should be limited to 5, i.e. $h \le 5b$. Substituting for $h = 5b$ in $I_{xx} = bh^3/12$ and equating it to $I_{xx.rqd}$:

   $I_{xx.rqd} = \dfrac{b \cdot (5 \cdot b)^3}{12}$ $\quad$ will give $\quad$ $b := \left(\dfrac{12 \cdot I_{xx.rqd}}{5^3}\right)^{1/4}$

   $\qquad\qquad\qquad\qquad\qquad\qquad$ $b = 56.64 \circ \mathrm{mm}$
   and depth of section, h $\qquad$ $h := 5 \cdot b$
   $\qquad\qquad\qquad\qquad\qquad\qquad$ $h = 283.18 \circ \mathrm{mm}$

Try a 70 mm × 288 mm section with 8 laminations of 36 mm in thickness

Trial beam dimensions:

Depth (mm), h $\qquad\qquad h := 288 \cdot mm$

Breadth (mm), b $\qquad\qquad b := 70 \cdot mm$

Modification factor K_7 for section depth:
to provide a dimensionless
value for K_7 in Mathcad, let: $h_1 := h \cdot mm^{-1}$

for $h < 300$ mm $\qquad K_7 := \left(\dfrac{300 \cdot mm}{h} \right)^{0.11}$

$\qquad\qquad\qquad\qquad K_7 = 1$

Glulam modification factors (Table 21):
BS 5268 : Part 2, Clause 3.2
Single-grade, 8 laminates in
C24 timber, by interpolation:

Bending // to grain, K_{15} $\qquad K_{15} := 1.40$

Compression perpendicular $\qquad K_{18} := 1.55$
to grain, K_{18}

Shear // to grain, K_{19} $\qquad K_{19} := 2.34$

Modulus of elasticity, K_{20} $\qquad K_{20} := 1.07$

5. *Bonding stress*

Applied bending moment $\qquad M := \dfrac{W_{med} \cdot L_e}{8}$

$\qquad\qquad\qquad\qquad M = 8.2 \circ kN \cdot m$

Section modulus, Z_{xx} $\qquad Z_{xx} := \dfrac{b \cdot h^2}{6}$

$\qquad\qquad\qquad\qquad Z_{xx} = 9.68 \times 10^5 \circ mm^3$

Applied bending stress $\qquad \sigma_{m.a.//} := \dfrac{M}{Z_{xx}}$

$\qquad\qquad\qquad\qquad \sigma_{m.a.//} = 8.47 \circ N \cdot mm^{-2}$

Permissible bending stress $\qquad \sigma_{m.adm.//} := \sigma_{m.g.//} \cdot K_2 \cdot K_3 \cdot K_6 \cdot K_7 \cdot K_8 \cdot K_{15}$

$\qquad\qquad\qquad\qquad \sigma_{m.adm.//} = 13.18 \circ N \cdot mm^{-2}$

Bending stress satisfactory

Check self-weight:
BS 5268 : Part 2, Table 7

Average density $\qquad \rho := 420 \cdot kg \cdot m^{-3}$

Beam self-weight, (kN/m)

Actual $\qquad\qquad Swt_{actual} := \rho \cdot g \cdot h \cdot b$

$\qquad\qquad\qquad\qquad Swt_{actual} = 0.08 \circ kN \cdot m^{-1}$

Assumed $\qquad\qquad Swt_{assumed} := Swt$

$\qquad\qquad\qquad\qquad Swt_{assumed} = 0.3 \circ kN \cdot m^{-1}$

Self-weight satisfactory

6. *Lateral stability*

Clause 2.10.8 and Table 16

Maximum depth-to-breadth ratio, h/b

$\dfrac{h}{b} = 4.11$

Ends should be held in position and compression edges held in line by direct connection of decking to the beams

7. *Shear stress*

Applied shear force

$$F_v := \frac{W_{med}}{2}$$

$F_v = 7.54 \circ \text{kN}$

Applied shear stress

$$\tau_{a.\parallel} := \frac{3}{2} \cdot \left(\frac{F_v}{b \cdot h} \right)$$

$\tau_{a.\parallel} = 0.56 \circ \text{N} \cdot \text{mm}^{-2}$

Permissible shear stress, no notch

$\tau_{adm.\parallel} := \tau_{g.\parallel} \cdot K_2 \cdot K_3 \cdot K_5 \cdot K_8 \cdot K_{19}$

$\tau_{adm.\parallel} = 2.08 \circ \text{N} \cdot \text{mm}^{-2}$

Shear stress satisfactory

8. *Bearing stress*

Applied load

$$F := \frac{W_{med}}{2}$$

$F = 7.54 \circ \text{kN}$

Applied bearing stress (bearing area $= b \cdot bw$)

$$\sigma_{c.a.pp} := \left(\frac{F}{b \cdot bw} \right)$$

$\sigma_{c.a.pp} = 1.08 \circ \text{N} \cdot \text{mm}^{-2}$

Permissible bearing stress

$\sigma_{c.adm.pp} := \sigma_{c.g.pp} \cdot K_2 \cdot K_3 \cdot K_4 \cdot K_8 \cdot K_{18}$

$\sigma_{c.adm.pp} = 3.68 \circ \text{N} \cdot \text{mm}^{-2}$

Bearing stress satisfactory

9. *Deflection*

Modulus of elasticity for glulam, E

$E := E_{mean} \cdot K_{20}$

$E = 1.16 \times 10^4 \circ \text{N} \cdot \text{mm}^{-2}$

Second moment of area, I_{xx}

$$I_{xx} := \frac{b \cdot h^3}{12}$$

$I_{xx} = 1.39 \times 10^8 \circ \text{mm}^4$

Deflection due to bending

$$\Delta_m := \frac{5 \cdot W_{med} \cdot L_e^3}{384 \cdot E \cdot I_{xx}}$$

$\Delta_m = 10.04 \circ \text{mm}$

Deflection due to shear

$$\Delta_s := \frac{19.2 \cdot M}{(b \cdot h) \cdot E}$$

$\Delta_s = 0.68 \circ \text{mm}$

Total deflection

$\Delta_{total} := \Delta_m + \Delta_s$

$\Delta_{total} = 10.71 \circ \text{mm}$

Permissible deflection

$\Delta_{adm} := 0.003 \cdot L_e$

$\Delta_{adm} = 13.05 \circ \text{mm}$

Deflection okay and no pre-cambering is
required

Therefore adopt 70 mm × 288 mm deep sections with 8 laminatins of 36 mm in thickness

C. *Design of main beams*

1. *Loading*

Effective length, L_e $L_e := 9.6 \cdot \text{m}$

Bearing width, bw $bw := 175 \cdot \text{mm}$

Beam dimensions:

Depth, h

breadth, b $\Bigg\}$ Not known at this stage

Self-weight of beam (kN/m), Swt $Swt := 0.7 \cdot \text{kN} \cdot \text{m}^{-1}$ assumed

Strip of load of width $b_j - 300 \text{ mm}$, see Fig. 6.8 (b), on the top of a main beam plus its
self-weight

$b_j := 300 \cdot \text{mm}$

Long-term loading $P_{strip.long} := (DL \cdot b_j + Swt) \cdot L_e$

$P_{strip.long} = 8.76 \circ \text{kN}$

Medium-term loading $P_{strip.med} := (DL + IL) \cdot b_j \cdot L_e + Swt \cdot L_e$

$P_{strip.med} = 18.12 \circ \text{kN}$

Point (concentrated) loading
from secondary beams (kN), P_s $P_{s.long} := 2 \cdot \dfrac{W_{long}}{2}$

$P_{s.long} = 3.77 \circ \text{kN}$

$P_{s.med} := 2 \cdot \dfrac{W_{med}}{2}$

$P_{s.med} = 15.08 \circ \text{kN}$

Total load (kN):

Long-term, P_{long} $P_{long} := (9 + 2 \cdot 0.5) \cdot P_{s.long} + P_{strip.long}$

$P_{long} = 46.47 \circ \text{kN}$

Medium-term, P_{med} $P_{med} := (9 + 2 \cdot 0.5) \cdot P_{s.med} + P_{strip.med}$

$P_{med} = 168.93 \circ \text{kN}$

2. *K-factors*

K_1 to K_6 are as above

No load sharing applies $K_8 := 1.0$
   (K_8, Clause 2.9)

Glulam modification factors (Table 21):
BS 5268 : Part 2, Clause 3.2
Bending // to grain, K_{15}
Compression perpendicular
to grain, K_{18}
Shear // to grain, K_{19} $\left.\rule{0pt}{5em}\right\}$ Since number of laminations is not known,
ignore at this stage
Modulus of elasticity, K_{20} $K_{20} := 1.07$

3. Grade stresses

As for the secondary beams (see above)

4. Selecting a trial section

(1) Using deflection criteria:
For a simply supported beam subjected to a maximum bending moment of M_{max}, irrespective of the loading type, the maximum deflection may be estimated using the equation given in Section 4.4.3, as:

$$\Delta_m = \frac{0.104 \cdot M_{max} \cdot L_e^2}{E \cdot I_{xx}}$$

M_{max} for a simply supported beam carrying $(n-1)$ point loads of magnitude P, equally spaced, is given by

$$M_{max} = \frac{n \cdot P \cdot L_e^2}{8} \qquad \text{(reference: \textit{Steel Designer's Manual})}$$

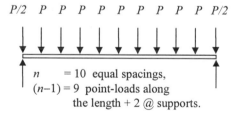

P/2	*P*	*P*	*P*	*P*	*P*	*P*	*P*	*P*	*P*	*P/2*

n = 10 equal spacings,
$(n-1)$ = 9 point-loads along
the length + 2 @ supports.

Number of point-loads, $n := 10$
$(n - 1) = 9$

Applied bending moment $M_p := \dfrac{n \cdot P_{s.med} \cdot L_e}{8}$
due to point loads
$M_p = 180.97 \circ \text{kN} \cdot \text{m}$

Applied bending moment $M_s := \dfrac{P_{strip.med} \cdot L_e}{8}$
due to strip of loading
including self-weight
$M_s = 21.74 \circ \text{kN} \cdot \text{m}$

Total bending moment $M_{total} := M_s + M_p$
$M_{total} = 202.71 \circ \text{kN} \cdot \text{m}$

Modulus of elasticity for glulam, E

$E := E_{mean} \cdot K_{20}$
$E = 1.16 \times 10^4 \circ \text{N} \cdot \text{mm}^{-2}$

Permissible deformation

$\Delta_{adm} := 0.003 \cdot L_e$
$\Delta_{adm} = 28.8 \circ \text{mm}$

Required second moment of area, I_{xx} using Δ_{adm}

$I_{xx.rqd} := \dfrac{0.104 \cdot M_{total} \cdot L_e^2}{E \cdot \Delta_{adm}}$

$I_{xx.rqd} = 5.84 \times 10^9 \circ \text{mm}^4$

(2) Using lateral stability criteria:
In order to achieve lateral stability by direct fixing of decking to beams, the depth-to-breadth ratio should be limited to 5, i.e. $h \leq 5b$. Substituting for $h = 5b$ in $I_{xx} = bh^3/12$ and equating it to $I_{xx,rqd}$:

$I_{xx,rqd} = \dfrac{b \cdot (5 \cdot b)^3}{12}$ will give $b := \left(\dfrac{12 \cdot I_{xx.rqd}}{5^3} \right)^{1/4}$

$b = 153.86 \circ \text{mm}$

and depth of the section, h

$h := 5 \cdot b$
$h = 769.31 \circ \text{mm}$

Try a 180 mm × 864 mm deep section with 24 laminations of 36 mm in thickness
Trial beam dimensions:

Depth (mm), h $h := 864 \cdot \text{mm}$

Breadth (mm), b $b := 180 \cdot \text{mm}$

Modification factor K_7 for section depth:
　To provide a dimensionless
　　value for K_7 in Mathcad,
　　let: $h_1 := h \cdot \text{mm}^{-1}$

　for $h > 300$ mm $K_7 := 0.81 \dfrac{h_1^2 + 92\,300}{h_1^2 + 56\,800}$

                          $K_7 = 0.85$

Glulam modification factors (Table 21):
BS 5268 : Part 2, Clause 3.2
Single-grade, 24 laminates in
  C24 timber

Bending // to grain, K_{15} $K_{15} := 1.52$

Compression perpendicular to grain, K_{18} $K_{18} := 1.55$

Shear // to grain, K_{19} $K_{19} := 2.34$

Modulus of elasticity, K_{20} $K_{20} := 1.07$

5. *Bending stress*

Total applied bending moment $M_{total} = 202.71 \circ \text{kN} \cdot \text{m}$

Section modulus, Z_{xx} $Z_{xx} := \dfrac{b \cdot h^2}{6}$

                               $Z_{xx} = 2.24 \times 10^7 \circ \text{mm}^3$

Applied bending stress

$$\sigma_{m.a.\parallel} := \frac{M_{total}}{Z_{xx}}$$

$$\sigma_{m.a.\parallel} = 9.05 \circ \text{N} \cdot \text{mm}^{-2}$$

Permissible bending stress

$$\sigma_{m.adm.\parallel} := \sigma_{m.g.\parallel} \cdot K_2 \cdot K_3 \cdot K_6 \cdot K_7 \cdot K_8 \cdot K_{15}$$

$$\sigma_{m.adm.\parallel} = 12.05 \circ \text{N} \cdot \text{mm}^{-2}$$

Bending stress satisfactory

Check self-weight:
BS 5268 : Part 2, Table 7
Average density
Beam self-weight, (kN/m)
Actual

$$\rho := 420 \cdot \text{kg} \cdot \text{m}^{-3}$$

$$Swt_{actual} := \rho \cdot g \cdot h \cdot b$$

$$Swt_{actual} = 0.64 \circ \text{kN} \cdot \text{m}^{-1}$$

$$Swt_{assumed} := Swt$$

Assumed

$$Swt_{assumed} = 0.7 \circ \text{kN} \cdot \text{m}^{-1}$$

Self-weight satisfactory

6. *Lateral stability*

Clause 2.10.8 and Table 16

Maximum depth-to-breadth ratio, *h/b*

$$\frac{h}{b} = 4.8$$

Ends should be held in position and compression edges held in line by direct connection of decking to the beams

7. *Shear stress*

Applied shear force

$$F_v := \frac{P_{med}}{2}$$

$$F_v = 84.46 \circ \text{kN}$$

Applied shear stress

$$\tau_{a.\parallel} := \frac{3}{2} \cdot \left(\frac{F_v}{b \cdot h} \right)$$

$$\tau_{a.\parallel} = 0.81 \circ \text{N} \cdot \text{mm}^{-2}$$

Permissible shear stress, no notch

$$\tau_{adm.\parallel} := \tau_{g.\parallel} \cdot K_2 \cdot K_3 \cdot K_5 \cdot K_8 \cdot K_{19}$$

$$\tau_{adm.\parallel} = 2.08 \circ \text{N} \cdot \text{mm}^{-2}$$

Shear stress satisfactory

8. *Bearing stress*

Applied load

$$F := \frac{P_{med}}{2}$$

$$F = 84.46 \circ \text{kN}$$

Applied bearing stress, (bearing area = *b* × *bw*)

$$\sigma_{c.a.pp} := \left(\frac{F}{b \cdot bw} \right)$$

$$\sigma_{c.a.pp} = 2.68 \circ \text{N} \cdot \text{mm}^{-2}$$

Permissible bearing stress

$$\sigma_{c.adm.pp} := \sigma_{c.g.pp} \cdot K_2 \cdot K_3 \cdot K_4 \cdot K_8 \cdot K_{18}$$
$$\sigma_{c.adm.pp} = 3.68 \circ N \cdot mm^{-2}$$

Bearing stress satisfactory

9. Deflection

Modulus of elasticity for
glulam, E

$$E := E_{mean} \cdot K_{20}$$
$$E = 1.16 \times 10^4 \circ N \cdot mm^{-2}$$

Second moment of area, I_{xx}

$$I_{xx} := \frac{b \cdot h^3}{12}$$
$$I_{xx} = 9.67 \times 10^9 \circ mm^4$$

Deflection due to bending

$$\Delta_m := \frac{0.104 \cdot M_{total} \cdot L_e^2}{E \cdot I_{xx}}$$
$$\Delta_m = 17.38 \circ mm$$

Deflection due to shear

$$\Delta_s := \frac{19.2 \cdot M_{total}}{(b \cdot h) \cdot E}$$
$$\Delta_s = 2.17 \circ mm$$

Total deflection

$$\Delta_{total} := \Delta_m + \Delta_s$$
$$\Delta_{total} = 19.54 \circ mm$$

Permissible deflection

$$\Delta_{adm} := 0.003 \cdot L_e$$
$$\Delta_{adm} = 28.8 \circ mm$$

Deflection okay and no pre-cambering is required

Therefore adopt 180 mm wide × 864 mm deep sections with 24 laminations of 36 mm in thickness

Example 6.4   Design of a curved glued laminated beam – selection of a suitable section size

Design of a curved glued laminated timber beam for the roof of a restaurant is required. The beam is to span 9.0 m centre to centre supported on 250 mm wide bearings under service class 1 conditions. It is proposed to use a single grade lay-up using softwood timber in strength class C18 with horizontal laminations of 30 mm finished thickness. The radius of curvature at mid-span bend is to be 6.5 m. The beam is subjected to a dead load of 1.65 kN/m, including self-weight, and an imposed medium-term load of 2.25 kN/m.

Definitions

Force, kN	$N :=$ newton
Length, m	$kN := 10^3 \cdot N$
Cross-sectional dimensions, mm	Direction parallel to grain, //
Stress, Nmm^{-2}	Direction perpendicular to grain, *pp*

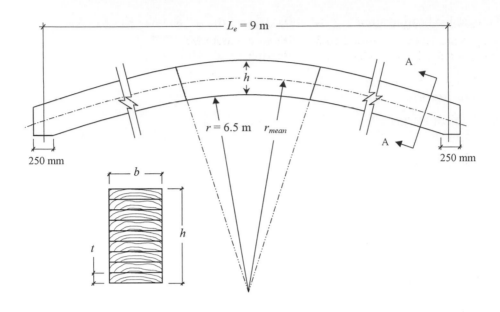

Cross-section A-A

Fig. 6.9 Curved glulam beam (Example 6.4).

1. *Loading*

Effective length, L_e	$L_e := 9.0 \cdot m$
Bearing width, bw	$bw := 250 \cdot mm$
Beam dimensions:	
Depth of section, h	$\Big\}$ Not known at this stage
Breadth of section, b	
Dead load including self-weight (kN/m), DL	$DL := 1.65 \cdot kN \cdot m^{-1}$
Imposed load (kN/m), IL	$IL := 2.25 \cdot kN \cdot m^{-1}$
Total load (kN):	
Long-term, W_{long}	$W_{long} := DL \cdot L_e$
	$W_{long} = 14.85 \circ kN$
Medium-term, W_{med}	$W_{med} := (DL + IL) \cdot L_e$
	$W_{med} = 35.1 \circ kN$

2. K-*factors*

Service class 1 (K_2, Table 13)	$K_2 := 1$
Medium-term loading (K_3, Table 14)	$K_3 := 1.25$
Bearing: 250 mm wide (K_4, Clause 2.10.2)	$K_4 := 1.0$
Notched end effect (K_5, Clause 2.10.4)	$K_5 := 1$

Form factor (K_6, Clause 2.10.5) $K_6 := 1$
Depth factor (K_7, Clause 2.10.6) $K_7 := 1$ assumed
No load sharing (K_8, Clause 2.9) $K_8 := 1$

Glulam modification factors (Table 21):
BS 5268 : Part 2, Clause 3.2
Bending // to grain, K_{15}
Compression perpendicular   } Since number of laminations is not known,
  to grain, K_{18} ignore at this stage
Shear // to grain, K_{19}
Modulus of elasticity, K_{20} $K_{20} := 1.17$

3. Grade stresses

BS 5268 : Part 2, Clause 3.2
Table 7, for C18 timber
Bending // to grain $\sigma_{m.g.//} := 5.8 \cdot N \cdot mm^{-2}$
Compression perpendicular $\sigma_{c.g.pp} := 1.7 \cdot N \cdot mm^{-2}$
  to grain
Shear // to grain $\tau_{g.//} := 0.67 \cdot N \cdot mm^{-2}$
Mean modulus of elasticity $E_{mean} := 9100 \cdot N \cdot mm^{-2}$

4. Selecting a trial section

(1) Using deflection criteria:
     Modulus of elasticity $E := E_{mean} \cdot K_{20}$
        for glulam, E $E = 1.06 \times 10^4 \circ N \cdot mm^{-2}$
     Permissible deformation $\Delta_{adm} := 0.003 \cdot L_e$
                                    $\Delta_{adm} = 27 \circ mm$

     Required second moment $I_{xx.rqd} := \dfrac{5 \cdot IL \cdot L_e^4}{384 \cdot E \cdot \Delta_{adm}}$
     of area, I_{xx} using Δ_{adm}
                                    $I_{xx.rqd} = 6.69 \times 10^8 \circ mm^4$

(2) Using lateral stability criteria:
     In order to achieve lateral stability by direct fixing of decking to beams, the depth-
     to-breadth ratio should be limited to 5, i.e. $h \leq 5b$. Substituting for $h = 5b$ in
     $I_{xx} = bh^3/12$ and equating it to $I_{xx.rqd}$:

$$I_{xx,rqd} = \frac{b \cdot (5 \cdot b)^3}{12} \quad \text{will give} \quad b := \left(\frac{12 \cdot I_{xx.rqd}}{5^3} \right)^{1/4}$$

                                    $b = 89.51 \circ mm$
     and depth of section, h $h := 5 \cdot b$
                                    $h = 447.55 \circ mm$

     Try a 105 mm × 510 mm deep section with 17 laminations of 30 mm in thickness
     Trial beam dimensions:
     Depth (mm), h $h := 510 \cdot mm$
     Breadth (mm), b $b := 105 \cdot mm$

Modification factor K_7 for section depth:
to provide a dimensionless
value for K_7 in Mathcad,
let: $h_1 := h \cdot \text{mm}^{-1}$

for $h > 300\,\text{mm}$ $K_7 := 0.81 \dfrac{h_1^2 + 92\,300}{h_1^2 + 56\,800}$

$K_7 = 0.9$

Glulam modification factors (Table 21):
BS 5268 : Part 2, Clause 3.2
Single-grade, 17 laminates in
  C18 timber by interpolation:
Bending // to grain $K_{15} := 1.52$
Compression perpendicular $K_{18} := 1.69$
  to grain
Shear // to grain $K_{19} := 2.73$
Modulus of elasticity $K_{20} := 1.17$

5. *Curvature*

BS 5268 : Part 2, Clause 3.5.3.1
Radius of curvature, r $r := 6500 \cdot \text{mm}$
Lamination thickness, t $t := 30 \cdot \text{mm}$

Ratio r/t should not be less $\dfrac{r}{t} = 216.67$ is greater than
  than $E_{mean}/70$

$\dfrac{E_{mean}}{70} = 130 \circ \text{N} \cdot \text{mm}^{-2}$

Satisfactory

6. *Bending stress*

BS 5268 : Part 2, Clause 3.5.3.2

Check if $(r/t) < 240$ $\dfrac{r}{t} = 216.67 \quad < 240$

hence $K_{33} := 0.76 + 0.001 \cdot \dfrac{r}{t}$

$K_{33} = 0.98$
<1.0 Satisfactory

Applied bending moment $M := \dfrac{W_{med} \cdot L_e}{8}$

$M = 39.49 \circ \text{kN} \cdot \text{m}$

Check if $(r_{mean}/h) < 15$ $r_{mean} := r + 0.5 \cdot h$
$r_{mean} = 6.75 \circ \text{m}$

$\dfrac{r_{mean}}{h} = 13.25 \quad <15$

Satisfactory

Since $10 < (r_{mean}/h) \le 15$

$$K_{34} := 1.15 - \left(0.01 \cdot \frac{r_{mean}}{h}\right)$$

$$K_{34} = 1.02$$

Section modulus

$$Z_{xx} := \frac{b \cdot h^2}{6}$$

$$Z_{xx} = 4.55 \times 10^6 \circ mm^3$$

Applied bending stress on concave face

$$\sigma_{m.a.concave} := K_{34} \cdot \frac{M}{Z_{xx}}$$

$$\sigma_{m.a.concave} = 8.83 \circ N \cdot mm^{-2}$$

Applied bending stress on convex face

$$\sigma_{m.a.convex} := \frac{M}{Z_{xx}}$$

$$\sigma_{m.a.convex} = 8.68 \circ N \cdot mm^{-2}$$

Permissible bending stress

$$\sigma_{m.adm.//} := \sigma_{m.g.//} \cdot K_2 \cdot K_3 \cdot K_6 \cdot K_7 \cdot K_8 \cdot K_{15} \cdot K_{33}$$
$$\sigma_{m.adm.//} = 9.69 \circ N \cdot mm^{-2}$$
Bending stress satisfactory

7. Radial stress

BS 5268 : Part 2, Clause 3.5.3.3

Radial stress, σ_r, induced by bending moment, M,

$$\sigma_r := \frac{3 \cdot M}{2 \cdot b \cdot h \cdot r_{mean}}$$

$$\sigma_r = 0.16 \circ N \cdot mm^{-2}$$

Since the applied moment, M, tends to increase the radius of curvature, σ_r is tension perpendicular to grain, and factor K_{19} is disregarded (Clause 3.2, BS 5268 : Part 2).

Permissible radial stress, $\sigma_{t.adm.pp} = \frac{1}{3} \cdot \tau_{adm.//}$ (see Section 2.5.7, Chapter 2)

therefore

$$\sigma_{r.adm} := \frac{1}{3} \cdot \tau_{g.//} \cdot K_2 \cdot K_3 \cdot K_8$$
$$\sigma_{r.adm} = 0.28 \circ N \cdot mm^{-2}$$
Radial stress satisfactory

8. Lateral stability

Clause 2.10.8 and Table 16

Maximum depth-to-breadth ratio, h/b

$$\frac{h}{b} = 4.86$$

Ends should be held in position and compression edges held in line by direct connection of decking to the beams

9. Shear stress

Applied shear force

$$F_v := \frac{W_{med}}{2}$$

$$F_v = 17.55 \circ kN$$

Applied shear stress

$$\tau_{a.\parallel} := \frac{3}{2} \cdot \left(\frac{F_v}{b \cdot h} \right)$$

$$\tau_{a.\parallel} = 0.49 \circ N \cdot mm^{-2}$$

Permissible shear stress, no notch

$$\tau_{adm.\parallel} := \tau_{g.\parallel} \cdot K_2 \cdot K_3 \cdot K_5 \cdot K_8 \cdot K_{19}$$

$$\tau_{adm.\parallel} = 2.29 \circ N \cdot mm^{-2}$$

Shear stress satisfactory

10. *Bearing stress*

Applied load

$$F_v := \frac{W_{med}}{2}$$

$$F_v = 17.55 \circ kN$$

Applied bearing stress
  (bearing area $= b \cdot bw$)

$$\sigma_{c.a.pp} := \left(\frac{F_v}{b.bw} \right)$$

$$\sigma_{c.a.pp} = 0.67 \circ N \cdot mm^{-2}$$

Permissible bearing stress

$$\sigma_{c.adm.pp} := \sigma_{c.g.pp} \cdot K_2 \cdot K_3 \cdot K_4 \cdot K_8 \cdot K_{18}$$

$$\sigma_{c.adm.pp} = 3.59 \circ N \cdot mm^{-2}$$

Bearing stress satisfactory.

11. *Deflection*

Modulus of elasticity for
  glulam, E

$$E := E_{mean} \cdot K_{20}$$

$$E = 1.06 \times 10^4 \circ N \cdot mm^{-2}$$

Second moment of area, I_{xx}

$$I_{xx} := \frac{b \cdot h^3}{12}$$

$$I_{xx} = 1.16 \times 10^9 \circ mm^4$$

Precamber = deflection
  under dead load

$$Camber := \frac{5 \cdot DL \cdot L_e^4}{384 \cdot E \cdot I_{xx}}$$

$$Camber = 11.41 \circ mm$$

Provide 12 mm camber

Deflection due to bending under
  imposed load

$$\Delta_m := \frac{5 \cdot IL \cdot L_e^4}{384 \cdot E \cdot I_{xx}}$$

$$\Delta_m = 15.55 \circ mm$$

Deflection due to shear

$$\Delta_s := \frac{19.2 \cdot M}{(b \cdot h) \cdot E}$$

$$\Delta_s = 1.33 \circ mm$$

Total deflection

$$\Delta_{total} := \Delta_m + \Delta_s$$

$$\Delta_{total} = 16.88 \circ mm$$

Permissible deflection

$$\Delta_{adm} := 0.003 \cdot L_e$$

$$\Delta_{adm} = 27 \circ mm$$

Deflection satisfactory.

Therefore adopt a 105 mm wide $\times$ 510 mm deep section with 17 laminations of 30 mm in thickness and 12 mm camber

Example 6.5   Load capacity of a glued laminated timber column

A single-grade glued laminated timber column in strength class C18 consists of 5 laminations of 37 mm thickness with overall cross-scctional dimensions of 97 mm × 185 mm.
The column is 4 m in height and is restrained at both ends in position but not in
direction and is subjected to service class 2 conditions.
(a)   Determine the maximum axial long-term load that the column can support.
(b)   Check that the section is adequate to resist a long-term axial load of 25 kN and a
bending moment of 1.40 kN/m due to lateral loading about its major axis.

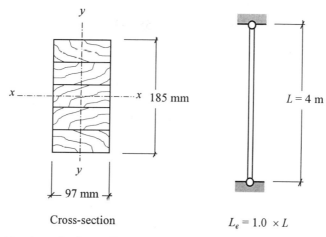

Cross-section $L_e = 1.0 \times L$

Fig. 6.10   Glued laminated column (Example 6 5)

Definitions

Force, kN $N := \text{newton}$
Length, m $kN := 10^3 \cdot N$
Cross-sectional dimensions, mm   Direction parallel to grain, //
Stress, Nmm^{-2} Direction perpendicular to grain, *pp*

1.   *Geometrical properties*

Column length, L $L := 4.0 \cdot m$
Table 18, effective length, L_e $L_e := 1.0 \cdot L$
                                   $L_e = 4 \circ m$
Width of section, b $b := 97 \cdot mm$
Depth of section, d $d := 185 \cdot mm$
Cross-sectional area, A $A := b \cdot d$
                                   $A = 1.79 \times 10^4 \circ mm^2$
Second moment of area, I_{xx} $I_{xx} := \frac{1}{12} \cdot b \cdot d^3$
                                   $I_{xx} = 5.12 \times 10^7 \circ mm^4$
Second moment of area, I_{yy} $I_{yy} := \frac{1}{12} \cdot d \cdot b^3$
                                   $I_{yy} = 1.41 \times 10^7 \circ mm^4$

For a rectangular section, radius of gyration, i_{xx}

$$i_{xx} := \frac{d}{\sqrt{12}}$$

$$i_{xx} = 53.4 \circ \text{mm}$$

and i_{yy}

$$i_{yy} := \frac{b}{\sqrt{12}}$$

$$i_{yy} = 28 \circ \text{mm}$$

Section modulus, Z_{xx}

$$Z_{xx} := \frac{b \cdot d^2}{6}$$

$$Z_{xx} = 5.53 \times 10^5 \circ \text{mm}^3$$

2. *Check slenderness ratio, λ*

About x–x axis

$$\lambda_{xx} := \frac{L_e}{i_{xx}}$$

$$\lambda_{xx} = 74.9$$

About y–y axis

$$\lambda_{yy} := \frac{L_e}{i_{yy}}$$

$$\lambda_{yy} = 142.85$$

<180 satisfactory

3. *Grade stresses*

BS 5268 : Part 2, Table 7
Strength class, C18 timber
Bending // to grain $\sigma_{m.g.//} := 5.8 \cdot \text{N} \cdot \text{mm}^{-2}$
Compression parallel to grain $\sigma_{c.g.//} := 7.1 \cdot \text{N} \cdot \text{mm}^{-2}$
Minimum modulus of elasticity $E_{min} := 6000 \cdot \text{N} \cdot \text{mm}^{-2}$
Mean modulus of elasticity $E_{mean} := 9100 \cdot \text{N} \cdot \text{mm}^{-2}$

4. *K-factors*

Service class 2 (K_2, Table 13) $K_2 := 1$
Long-term loading (K_3, Table 14) $K_3 := 1$
Form factor (K_6, Clause 2.10.5) $K_6 := 1$

Depth factor (K_7, Clause 2.10.6) $K_7 := \left(\frac{300 \cdot \text{mm}}{d} \right)^{0.11}$

$$K_7 = 1.05$$

No load sharing (K_8, Clauses 2.9 and 2.11.5) $K_8 := 1$

Part (a)

Glulam modification factors:
Since buckling is critical about column's minor axis, i.e. y–y axis, the behaviour under axial loading only is similar to the vertically laminated members and hence the

modification factor K_{28}, given in Table 22 for compression // to grain and modulus of elasticity, applies.

Single-grade 5 laminates in
  C18 softwood timber:
Compression // to grain and $K_{28} := 1.27$
  modulus of elasticity

Modification factor for compression member, K_{12}, can be calculated using either one of the following methods:

(1) Using the equation in Annex B:
      BS 5268 : Part 2, Clause 3.3 $E := E_{min} \cdot K_{28}$
$$E = 7.62 \times 10^3 \circ N \cdot mm^{-2}$$

$$\lambda := \lambda_{yy} \quad \sigma_c := \sigma_{c.g.//} \cdot K_2 \cdot K_3 \quad \text{and} \quad \eta := 0.005 \cdot \lambda$$

$$K_{12} := \frac{1}{2} + \frac{(1+\eta) \cdot \pi^2 \cdot E}{3 \cdot \lambda^2 \cdot \sigma_c} - \sqrt{\left(\frac{1}{2} + \frac{(1+\eta) \cdot \pi^2 \cdot E}{3 \cdot \lambda^2 \cdot \sigma_c}\right)^2 - \frac{\pi^2 \cdot E}{1.5 \cdot \lambda^2 \cdot \sigma_c}}$$

$$K_{12} = 0.26$$

(2) Using Table 19:

$$\frac{E}{\sigma_{c.g.//} \cdot K_2 \cdot K_3} = 1.07 \times 10^3 \quad \text{and} \quad \lambda = 142.85 \quad K_{12} = 0.26$$

5. *Permissible compressive stress*

$$\sigma_{c.adm.//} := \sigma_{c.g.//} \cdot K_2 \cdot K_3 \cdot K_8 \cdot K_{12} \cdot K_{28}$$
$$\sigma_{c.adm.//} = 2.34 \circ N \cdot mm^{-2}$$

Hence the axial long-term load $P_{capacity} := \sigma_{c.adm.//} \cdot A$
  capacity of the column is: $P_{capacity} = 41.98 \circ kN$

Part (b)

6. *Applied loading*

Axial load $P := 25 \cdot kN$
Applied moment $M := 1.40 \cdot kN \cdot m$

Glulam modification factors (Table 21):
Since bending is about column's major axis, i.e. x–x axis, the behaviour is similar to horizontally laminated beams and hence modification factors K_{15}–K_{20} apply.

Single-grade 5 laminates in
  C18 softwood timber
Bending // to grain $K_{15} := 1.16$
Compression // to grain $K_{17} := 1.07$
Modulus of elasticity $K_{20} := 1.17$

Modification factor for compression member, K_{12}, can be calculated using either one of the following methods:

(1) Using the equation in Annex B:

BS 5268 : Part 2, Clause 3.6 $E := E_{mean} \cdot K_{20}$

$$E = 1.06 \times 10^4 \circ N \cdot mm^{-2}$$

$$\lambda := \lambda_{yy} \quad \sigma_c := \sigma_{c.g.\parallel} \cdot K_2 \cdot K_3 \quad \text{and} \quad \eta := 0.005 \cdot \lambda$$

$$K_{12} := \frac{1}{2} + \frac{(1+\eta) \cdot \pi^2 \cdot E}{3 \cdot \lambda^2 \cdot \sigma_c} - \sqrt{\left(\frac{1}{2} + \frac{(1+\eta) \cdot \pi^2 \cdot E}{3 \cdot \lambda^2 \cdot \sigma_c}\right)^2 - \frac{\pi^2 \cdot E}{1.5 \cdot \lambda^2 \cdot \sigma_c}}$$

$$K_{12} = 0.32$$

(2) Using Table 19:

$$\frac{E}{\sigma_{c.g.\parallel} \cdot K_2 \cdot K_3} = 1.5 \times 10^3 \quad \text{and} \quad \lambda = 142.85 \quad K_{12} = 0.32$$

7. Compression and bending stresses

Applied compressive stress $\sigma_{c.a.\parallel} := \dfrac{P}{A}$

$$\sigma_{c.a.\parallel} = 1.39 \circ N \cdot mm^{-2}$$

Permissible compressive stress $\sigma_{c.adm.\parallel} := \sigma_{c.g.\parallel} \cdot K_2 \cdot K_3 \cdot K_8 \cdot K_{12} \cdot K_{17}$

$$\sigma_{c.adm.\parallel} := 2.44 \circ N \cdot mm^{-2}$$

Compressive stress satisfactory

Applied bending stress $\sigma_{m.a.\parallel} := \dfrac{M}{Z_{xx}}$

$$\sigma_{m.a.\parallel} = 2.53 \circ N \cdot mm^{-2}$$

Permissible bending stress $\sigma_{m.adm.\parallel} := \sigma_{m.g.\parallel} \cdot K_2 \cdot K_3 \cdot K_6 \cdot K_7 \cdot K_8 \cdot K_{15}$

$$\sigma_{m.adm.\parallel} = 7.1 \circ N \cdot mm^{-2}$$

Bending stress satisfactory

Check interaction quantity

Euler critical stress $\sigma_e := \dfrac{\pi^2 \cdot E_{mean} \cdot K_{20}}{\lambda^2}$

$$\sigma_e = 5.15 \circ N \cdot mm^{-2}$$

$$\frac{\sigma_{m.a.\parallel}}{\sigma_{m.adm.\parallel} \cdot \left(1 - \dfrac{1.5 \cdot \sigma_{c.a.\parallel}}{\sigma_e} \cdot K_{12}\right)} + \frac{\sigma_{c.a.\parallel}}{\sigma_{c.adm.\parallel}}$$

$$= 0.98 < 1$$

Interaction quantity is satisfactory

Therefore a 97 mm × 185 mm section with 5 laminations of 37 mm timber in strength class C18 is satisfactory

Chapter 7
Design of Ply-webbed Beams

7.1 Introduction

A ply-webbed beam comprises three main parts, namely flanges, web(s) and glued or nailed joints between flanges and web(s). The flanges are generally made of solid structural timber sections finger-jointed together to produce the required length. The web(s) are made of plywood, although other wood-based materials such as particleboard or fibreboard may also be used. The flanges are often connected to plywood web(s) by structural timber adhesives in an industrial process, or sometimes by mechanical fasteners such as nails.

Ply-webbed beams are usually of I or boxed shapes. Typical cross-sections of ply-box and I beams which are symmetrical about both vertical and horizontal axes are shown in Fig. 7.1.

Ply-webbed beams owe their structural advantage to plywood acting as a shear-resistant material for the webs, allowing the solid timber sections to be spaced apart. This permits saving in materials and provides extra stiffness and strength for the beams. Ply-webbed beams are widely used for spans in excess of those for which solid timber is suitable, i.e. for spans greater than

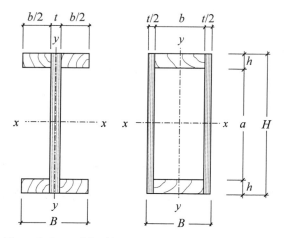

Fig. 7.1 I and boxed ply-webbed beams.

5 m and often up to 10–15 m. Their commonest application is in roof construction, where parallel top and bottom flanges may be used to give a flat roof or the upper profile may be pitched[1].

Ply-webbed beams have also been used as a type of large span (30 m or more) built-up portal frame in bridge construction. In addition, the good strength to weight performance of ply-webbed beams makes them an attractive option for construction of large cross-sectioned beams. For example as a hollow interior boxed bridge construction for a covered pedestrian crossing or as a prefabricated housing unit[1].

In this chapter, design recommendations for the most common method of construction, i.e. glued ply-webbed beams, are discussed. The recommendations may also be applied, with some modification, to all nailed ply-webbed beams.

7.2 Transformed (effective) geometrical properties

In general a ply-webbed beam is built with materials which have different moduli of elasticity, which lead to development of different stress magnitudes at the same depth. In Fig. 7.2 an example of differences in stress values for the two beam profiles subjected to bending moment is illustrated.

For a ply-webbed beam comprising materials with different moduli of elasticity, it is common practice to calculate the *effective geometrical properties* of the cross-section using a technique known as *transformed section method*. In this method the whole cross-section is considered as acting as one homogeneous material with the same properties as say the flange material, where the contribution from the web(s) is altered in proportion to the ratio of the moduli of elasticity.

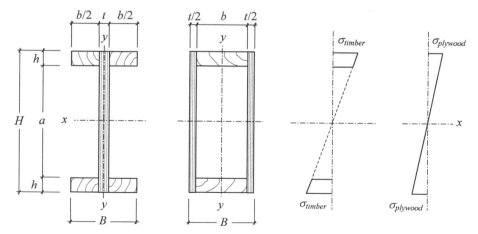

Fig. 7.2   Bending stresses in ply-webbed beams.

Therefore if it is required, for example, to transform a section to an equivalent of all timber, i.e. by replacing the plywood web with an equivalent thickness (width) of timber, then:

$$t_{timber} = t_{plywood} \times \frac{E_{plywood}}{E_{timber}}$$ (7.1a)

Transformed or effective area may be calculated as:

$$A_{transformed} = A_{timber(flanges)} + A_{plywood(webs)} \times \frac{E_{plywood}}{E_{timber}}$$ (7.1b)

and transformed or effective second moment of area as:

$$I_{transformed} = I_{timber(flanges)} + I_{plywood(webs)} \times \frac{E_{plywood}}{E_{timber}}$$ (7.1c)

Note: In calculation of transformed (effective) geometrical properties of a ply-webbed cross-section, extensive research and experience in practice have shown that the use of E_{mean} for timber is more realistic.[1]

7.3 Plywood

Before discussing design considerations for the ply-webbed beams it may be appropriate to describe briefly the main properties, particularly the structural behaviour, of plywood.

Plywood is a panel made up of a number of softwood veneers. The veneers are glued together so that the grain of each layer is perpendicular to the grain of its adjoining layer. Plywood as a structural material consists of an odd number of layers or plies (at least three) as shown in Fig. 7.3. Plywood for exterior use is generally made with fully waterproof adhesive that is suitable for severe exposure conditions.

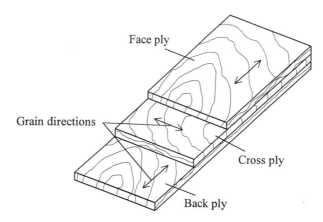

Fig. 7.3 The structure of a three-ply plywood.

The structural properties and strength of plywood depend mainly on the number and thickness of each ply, the species and grade, and the arrangement of the individual plies. As with timber, load factors such as type of applied stresses, their direction with respect to grain direction of face ply and duration of loading have great influence on the structural properties of plywood.

Plywood may be subjected to bending in two different planes depending on its intended use and the direction of the applied stress and therefore it is important to differentiate between them:

(1)   Bending about either of the axes (i.e. x–x or y–y) in the plane of the board, as shown in Fig. 7.4(a), for example in situations where it is used as shelving or as floorboard.
(2)   Bending about an axis perpendicular to the plane of the panel (i.e. z–z), as shown in Fig. 7.4(b), for example when it is acting as a web of a flexural member such as in ply-webbed beams.

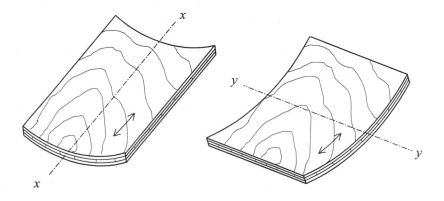

(a) Bending about either of the axes in the plane of the board.

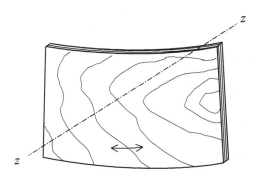

(b) Bending about an axis perpendicular to the plane of the board.

Fig. 7.4   Plywood – axes of bending.

Table 7.1 Modification factor K_{36} for the grade stresses and moduli of plywood (Table 33 of BS 5268 : Part 2)

Duration of loading	Value of K_{36}			
	Service classes 1 and 2		Service class 3	
	Stress	Modulus	Stress	Modulus
Long term	1.00	1.00	0.83	0.57
Medium term	1.33	1.54	1.08	1.06
Short and very short term	1.50	2.00	1.17	1.43

BS 5268 : Part 2:1996, Section 4 deals with plywood for structural use. Dimensions and section properties of plywoods are given in Tables 25–32 of the code. They are based on the minimum thicknesses permitted by the relevant product standards and apply to all service classes.

The grade stresses and moduli for plywoods are given in Tables 34–47 of the code. These apply to long-duration loading in service classes 1 and 2, and should be used in conjunction with the corresponding section properties given in Tables 25–32. For other durations of load and/or service class 3 condition, the stresses and moduli should be multiplied by the modification factor K_{36} given in Table 33 of the code. A summary of this table is reproduced here as Table 7.1.

It is important to note that the bending stresses and moduli given in Tables 34–47 of the code apply when the bending is perpendicular to the plane of the plywood panel (i.e. case (1) above). In situations where plywood is subject to bending about an axis perpendicular to the plane of the board (i.e. with the edge loaded, as in a ply-webbed beam) the tensile and compressive stresses induced by the bending moment should be considered individually, and the tension and compression stresses and moduli for the appropriate face grain orientation should be used.

7.4 Design considerations

The following criteria should be considered when designing ply-web beams:

(1) Bending.
(2) Deflection.
(3) Panel shear.
(4) Rolling shear.
(5) Lateral stability.
(6) Web-stiffeners.

7.4.1 *Bending*

BS 5268 : Part 2, Clause 4.6 deals with the use of plywood in flexural members. The permissible stresses for plywood in flexural members are governed by the particular conditions of service and loading described in Clauses 2.8 and 2.9 of the code (which relate to load duration and load sharing respectively) and should be taken as the product of the grade stresses given in Tables 34–47 and the modification factor K_{36} from Table 33 of the code (Table 7.1).

BS 5268 : Part 2, Clause 2.10.10 specifies that the modification factors K_{27}, K_{28} and K_{29} (for glulam members) given in Table 22 may also be used for the flanges of glued built-up beams such as I and box ply-webbed beams. The number of pieces of timber in each flange should be taken as the number of laminations, irrespective of their orientation, to determine the value of the stress modification factors K_{27}, K_{28} and K_{29} for that flange.

As mentioned earlier, the plywood web(s) of a ply-webbed beam will be subjected to bending about an axis perpendicular to the plane of the board; and it is therefore necessary to check that the maximum applied bending stress in the transformed section is not greater than the lesser of:

(1) the permissible tensile stress of the timber flange, $\sigma_{timber,t,adm,/\!/}$
(2) the permissible compressive stress of the timber flange, $\sigma_{timber,c,adm,/\!/}$
(3) the permissible transformed tensile stress of the plywood web,

$$\sigma_{plywood,t,adm,/\!/} \cdot \frac{E_{plywood}}{E_{timber}}$$

and

(4) the permissible transformed compressive stress of the plywood web,

$$\sigma_{plywood,c,adm,/\!/} \cdot \frac{E_{plywood}}{E_{timber}}$$

Therefore the applied bending stress in the transformed section, $\sigma_{m,a,/\!/}$, should not be greater than the lesser of the above permissible stresses, where,

the permissible tensile stress for timber flanges is

$$\sigma_{timber,t,adm,/\!/} = \text{grade stress (Table 7), } \sigma_{t,g,/\!/} \times K_2 K_3 K_8 K_{14} K_{27} \qquad (7.2a)$$

the permissible compressive stress for timber flanges is

$$\sigma_{timber,c,adm,/\!/} = \text{grade stress (Table 7), } \sigma_{c,g,/\!/} \times K_2 K_3 K_8 K_{28} \qquad (7.2b)$$

and for plywood web(s), the permissible tensile or compressive stress is

$$\sigma_{plywood,t\,or\,c,adm,/\!/} = \text{grade stress (Tables 34–47), } \sigma_{plywood,t\,or\,c,g,/\!/}$$

$$\times \frac{E_{plywood}}{E_{timber}} K_8 K_{36} \qquad (7.2c \ \& \ d)$$

The applied bending stress for a beam subjected to a maximum bending moment of M can be calculated from:

$$\sigma_{m,a,\parallel} = \frac{M}{Z_{xx}} \qquad (7.3)$$

where: $Z_{xx} = \dfrac{I_{xx}}{y}$ $y = \dfrac{H}{2}$ and $I_{xx} = \dfrac{BH^3}{12} - \dfrac{ba^3}{12}$

a, b, B and H are as indicated in Fig. 7.1.

7.4.2   *Deflection*

Generally, in composite constructions, such as ply-webbed beams, the plane cross-sections of the beam do not remain plane during bending, as assumed in simple theory. This leads to increased deformation due to shear. Therefore deflection of a ply-webbed beam, as mentioned in Section 4.4, is determined as the sum of deflections due to bending, Δ_m, and shear, Δ_s:

$$\Delta_{total} = \Delta_m + \Delta_s \qquad (7.4)$$

For deflection calculations Clause 2.10.10 of the code also states that the total number of pieces of timber in both flanges should be taken as the number of laminations necessary to determine the value of K_{28} that is to be applied to the minimum modulus of elasticity, E_{min}.

(1)   *Bending deflection*, Δ_m, for a simply supported beam carrying a uniformly distributed load of w/m, may be calculated from:

$$\Delta_m = \frac{5wL^4}{384EI} \qquad (7.5)$$

and for a simply supported beam carrying a concentrated load of P at mid-span, from:

$$\Delta_m = \frac{PL^3}{48EI} \qquad (7.6)$$

where:
$E = E_{min}$ (for timber) $\times K_{28}$ (for the total number of flanges in the beam, Table 22)
$I = $ second moment of area of the transformed section.

(2)   *Shear deflection*, Δ_s. There are many methods available to estimate the shear deflection in ply-webbed beams. However, for a simply supported beam, based on Roark's recommended approximation (1956),[2] the shear deflection may be estimated from:

$$\Delta_s = \frac{M}{G_W A_W} \qquad (7.7)$$

where:

M = the bending moment at mid-span

G_W = the modulus of rigidity (shear modulus) of plywood web

A_W = the area of the plywood web (actual, not transformed).

7.4.3 Panel shear

Applied panel shear, τ_a, is the horizontal shear stress induced in a plywood web and can be calculated from:

$$\tau_a = \frac{F_v Q}{I \cdot t} \tag{7.8}$$

where:

F_v = maximum shear force (usually maximum reaction)

Q = first moment about the neutral axis of the transformed section area above the axis

I = second moment of area of the whole transformed section

t = total thickness of the plywood (actual, not transformed).

Applied panel shear, τ_a, should not be greater than the permissible panel shear for plywood web, $\tau_{plywood,adm}$:

$$\tau_a \leq \tau_{plywood,adm}$$

where:

$$\tau_{plywood,adm} = \text{plywood grade stress for shear, } \tau_{plywood,g} \times K_8 K_{36} \tag{7.9}$$

7.4.4 Rolling shear

The term *rolling shear* is used to describe a failure mode which may result due to overstressing of the glue line between the flange and web of a beam by shear force(s) acting at the interface. This is due to the physical construction of plywood (i.e. having ply grains at 90° to each other), which enables the material fibres to roll across each other, creating a shear failure plane. The rolling shear stress will occur in veneer which has vertical grain. If face grain is vertical, the rolling shear will occur in the veneer that is glued to the flange. If the face grain is parallel to the flange, it will occur in the second veneer or the first glue line. In order to avoid rolling action taking place, the rolling shear must be limited by providing sufficient glued contact depth between the flange and web.

A simplified equation for calculation of applied rolling shear stress, $\tau_{r,a}$, is given by:

$$\tau_{r,a} = \frac{F_v \cdot Q_r}{2h \cdot I} \tag{7.10}$$

where:

F_v = maximum shear force (usually maximum reaction)
Q_r = first moment of area of upper flange about neutral axis
h = contact depth (depth of flange)
I = second moment of area of whole transformed section.

BS 5268 : Part 2, Clause 4.6 recommends that for rolling shear at the junction of the web and flange of a ply-webbed beam, and at the junction of the outmost longitudinal member, the grade stresses given in Tables 34–47 should be multiplied by the stress concentration modification factor, K_{37}, which has the value 0.5.

Applied rolling shear stress, $\tau_{r,a}$, should be less than or equal to permissible rolling shear stress

$$\tau_{r,a} \leq \tau_{r,adm}$$

where:

$$\tau_{r,adm} = \text{grade stress, } \tau_{r,g} \times K_8 K_{36} K_{37} \quad \text{and} \quad K_{37} = 0.5 \tag{7.11}$$

7.4.5 *Lateral stability*

Lateral stability of I and box beams should be checked with reference to Clause 2.10.10 of BS 5268 : Part 2. A summary of this clause is given below.

The ratio of the second moment of area of the cross-section about the neutral axis, I_{xx}, to the second moment of area about the axis perpendicular to the neutral axis, I_{yy}, should be checked to ensure that there is no risk of buckling under design load. This is achieved by following the code's recommendations as summarised in Table 7.2.

7.4.6 *Web-stiffeners*

Clause 2.10.10 of BS 5268 : Part 2 recommends that built-up members with thin webs, i.e. I and boxed ply-webbed beams, should be provided with

Table 7.2 Recommendations for lateral support for ply-webbed beams (Clause 2.10.10, BS 5268 : Part 2)

Degree of lateral support	Ratio I_{xx}/I_{yy}
(a) No lateral support is necessary	5 or less
(b) Ends of the beam should be held in position at the bottom flange at the supports	between 5 and 10
(c) The beam should be held in line at the ends	between 10 and 20
(d) One edge should be held in line	between 20 and 30
(e) The beam should be restrained by bridging or other bracing at intervals of not more than 2.4 m	between 30 and 40
(f) The compression flange should be fully restrained	greater than 40

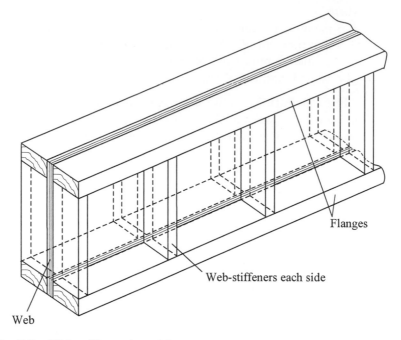

Fig. 7.5   Web-stiffeners in an I-beam.

web-stiffeners at all points of concentrated load including support positions or elsewhere as may be necessary.

Since there is no definite guidance available on the positioning of web-stiffeners along the length of the beam, the manufacturers of ply-webbed beams often have to revert to tests to prove the chosen centres, or adopt a conservative approach by placing stiffeners closer than probably required. In Fig. 7.5 web-stiffeners in an I-beam are illustrated.

7.5   References

1. Mettem, C. J. (1986) *Structural Timber Design and Technology*. Longman Scientific and Technical, Harlow.
2. Roark, R. J. (1956) *Formulas for Stress and Strain*. 4th edn. McGraw-Hill, New York.

7.6   Design examples

Example 7.1   Design of a ply-webbed box beam

Show that a glued ply-webbed box-beam with cross-sectional details as indicated in Fig. 7.6 is suitable for carrying a long-term uniformly distributed load of 1.95 kN/m including self-weight over an effective span of 6.5 m in service class 2.

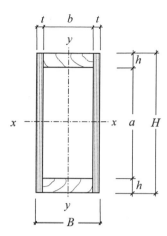

Fig. 7.6 Box ply-webbed beam (Example 7.1).

Design data

Flanges: Timber in strength
  class C16
Webs: Finnish birch, 9-plies,
  12 mm thickness with face
  grain // to span

Cross-sectional details:
  $b = 120$ mm
  $t = 12$ mm
  $h = 60$ mm
  $H = 500$ mm

Definitions

Force, kN
Length, m
Cross-sectional dimensions, mm
Stress, Nmm^{-2}

$N := $ newton
$kN := 10^3 \cdot N$
Direction parallel to grain, //
Direction perpendicular to grain, *pp*

1. Loading

Effective length, L_e
Long-term load (kN/m), w

$L_e := 6.5 \cdot m$
$w := 1.95 \cdot kN \cdot m^{-1}$

2. K-factors

Service class 2 (K_2, Table 13)
Long-term loading (K_3, Table 14)
No load sharing (K$_8$, Clause 2.9)
Width factor, tension member
  (K_{14}, Clause 2.12.2)

$K_2 := 1$
$K_3 := 1$
$K_8 := 1$
$b := 120 \cdot mm$

$$K_{14} := \left(\frac{300 \cdot mm}{b} \right)^{0.11}$$

$K_{14} = 1.11$

Clause 2.10.10

Tension //, (K_{27}, Table 22) $K_{27} := 1$ for one glued timber section per flange

Compression //, (K_{28c}, Table 22) $K_{28c} := 1$ for one glued timber section per flange

Modulus of elasticity for deflection (K_{28E}, Table 22) $K_{28E} := 1.14$ for a total of two glued timber sections in both flanges

Clause 4.5

Plywood grade stresses (K_{36s}, Table 33) $K_{36s} := 1.0$

Plywood moduli, (K_{36E}, Table 33) $K_{36E} := 1.0$

Plywood rolling shear, (K_{37}, Clause 4.6) $K_{37} := 0.5$

3.   Grade stresses

BS 5268 : Part 2, Table 7

Strength class, C16 timber

Tension // to grain $\sigma_{timber.t.g.//} := 3.2 \cdot N \cdot mm^{-2}$

Compression // to grain $\sigma_{timber.c.g.//} := 6.8 \cdot N \cdot mm^{-2}$

Shear // to grain $\tau_{timber.g.//} := 0.67 \cdot N \cdot mm^{-2}$

Mean E value $E_{timber.mean} := 8800 \cdot N \cdot mm^{-2}$

Minimum E value $E_{timber.min} := 5800 \cdot N \cdot mm^{-2}$

BS 5268 : Part 2, Table 40

Finnish birch, 9-plies, 12 mm plywood

Tension // to face grain $\sigma_{ply.t.g.//} := 19.16 \cdot N \cdot mm^{-2}$

Compression // to face grain $\sigma_{ply.c.g.//} := 9.8 \cdot N \cdot mm^{-2}$

Panel shear // and perpendicular to face grain $\tau_{ply.g} := 4.83 \cdot N \cdot mm^{-2}$

Rolling shear $\tau_{r.g} := 1.23 \cdot N \cdot mm^{-2}$

Modulus of elasticity in tension and compression // to face grain $E_{ply} := 4250 \cdot N \cdot mm^{-2}$

Modulus of rigidity (shear modulus) $G_w := 320 \cdot N \cdot mm^{-2}$

4.   Geometrical properties

Transform section to an equivalent of all timber.

Using E_{mean} for timber $E_{timber} := E_{timber.mean}$

Plywood thickness $t_{ply} := 12 \cdot mm$

Transformed web thickness, $t_{ply.tfd}$ $t_{ply.tfd} := t_{ply} \cdot \dfrac{E_{ply}}{E_{timber}}$

$t_{ply.tfd} = 5.8 \circ mm$

Transformed section:
Cross-sectional dimensions:

$b := 120 \cdot \text{mm}$

$B := 2 \cdot t_{ply.tfd} + b$

$B = 131.59 \circ \text{mm}$

$H := 500 \cdot \text{mm}$

$h := 60 \cdot \text{mm}$

$a := H - 2 \cdot h$

$a = 380 \circ \text{mm}$

Second moment of area, I_{xx}

$$I_{xx} := \frac{B \cdot H^3}{12} - \frac{b \cdot a^3}{12}$$

$I_{xx} = 8.22 \times 10^8 \circ \text{mm}^4$

Section modulus, Z_{xx}

$$Z_{xx} := \frac{I_{xx}}{\left(\dfrac{H}{2}\right)}$$

$Z_{xx} = 3.29 \times 10^6 \circ \text{mm}^3$

Second moment of area, I_{yy}

$$I_{yy} := \frac{H \cdot B^3}{12} - \frac{a \cdot b^3}{12}$$

$I_{yy} = 4.02 \times 10^7 \circ \text{mm}^4$

5. *Bending stress*

The maximum applied bending stress in transformed section should not be greater than the lesser of:

(1) Timber tension // to grain

$\sigma_{timber.t.adm.//} := \sigma_{timber.t.g.//} \cdot K_2 \cdot K_3 \cdot K_8 \cdot K_{14} \cdot K_{27}$

$\sigma_{timber.t.adm.//} = 3.54 \circ \text{N} \cdot \text{mm}^{-2}$

(2) Timber compression // to grain

$\sigma_{timber.c.adm.//} := \sigma_{timber.c.g.//} \cdot K_2 \cdot K_3 \cdot K_8 \cdot K_{28c}$

$\sigma_{timber.c.adm.//} = 6.8 \circ \text{N} \cdot \text{mm}^{-2}$

(3) Transformed plywood tension // to face grain

$\sigma_{tfd.ply.t.adm.//} := \sigma_{ply.t.g.//} \cdot \dfrac{E_{ply}}{E_{timber}} \cdot K_8 \cdot K_{36s}$

$\sigma_{tfd.ply.t.adm.//} = 9.25 \circ \text{N} \cdot \text{mm}^{-2}$

(4) Transformed plywood compression // to face grain

$\sigma_{tfd.ply.c.adm.//} := \sigma_{ply.c.g.//} \cdot \dfrac{E_{ply}}{E_{timber}} \cdot K_8 \cdot K_{36s}$

$\sigma_{tfd.ply.c.adm.//} = 4.73 \circ \text{N} \cdot \text{mm}^{-2}$

Thus, the **lowest** value is taken as the permissible stress in bending

$\sigma_{timber.t.adm.//} = 3.54 \circ \text{N} \cdot \text{mm}^{-2}$

Applied load, w

$w = 1.95 \circ \text{kN} \cdot \text{m}^{-1}$

Applied bending moment, M

$$M := \frac{w \cdot L_e^2}{8}$$

$M = 10.3 \circ \text{kN} \cdot \text{m}$

Applied bending stress, $\sigma_{m.a.\parallel}$

$$\sigma_{m.a.\parallel} := \frac{M}{Z_{xx}}$$

$$\sigma_{m.a.\parallel} = 3.13 \circ \text{N} \cdot \text{mm}^{-2}$$
Bending stress satisfactory

6. Deflection

BS 5268 : Part 2, Clause 2.10.10

Modulus of elasticity for deflection

$$E := E_{timber.min} \cdot K_{28E}$$
$$E = 6.61 \times 10^3 \circ \text{N} \cdot \text{mm}^{-2}$$

Deflection due to bending

$$\Delta_m := \frac{5 \cdot w \cdot L_e^4}{384 \cdot E \cdot I_{xx}}$$
$$\Delta_m = 8.34 \circ \text{mm}$$

Web area, A_w

$$A_w := 2 \cdot (H \cdot t_{ply})$$
$$A_w = 1.2 \times 10^4 \circ \text{mm}^2$$

Deflection due to shear

$$\Delta_s := \frac{M}{G_w \cdot A_w}$$
$$\Delta_s = 2.68 \circ \text{mm}$$

Total deflection

$$\Delta_{total} := \Delta_m + \Delta_s$$
$$\Delta_{total} = 11.02 \circ \text{mm}$$

Permissible deflection

$$\Delta_{adm} := 0.003 \cdot L_e$$
$$\Delta_{adm} = 19.5 \circ \text{mm}$$
Deflection satisfactory

7. Panel shear

Applied shear force, F_v

$$F_v := \frac{w \cdot L_e}{2}$$
$$F_v = 6.34 \circ \text{kN}$$

First moment of area, Q

$$Q := (b \cdot h) \cdot \left(\frac{H}{2} - \frac{h}{2}\right) + 2 \cdot \left(t_{ply.tfd} \cdot \frac{H}{2} \cdot \frac{H}{4}\right)$$
$$Q = 1.95 \times 10^6 \circ \text{mm}^3$$

Applied panel shear stress, τ_a

$$\tau_a := \frac{F_v \cdot Q}{I_{xx} \cdot (2 \cdot t_{ply})}$$
$$\tau_a = 0.63 \circ \text{N} \cdot \text{mm}^{-2}$$

Permissible panel shear

$$\tau_{adm.\parallel} := \tau_{ply.g} \cdot K_8 \cdot K_{36s}$$
$$\tau_{adm.\parallel} = 4.83 \circ \text{N} \cdot \text{mm}^{-2}$$
Panel shear satisfactory

8. Rolling shear

Applied shear force, F_v

$$F_v = 6.34 \circ \text{kN}$$

First moment of area of upper flange, Q_r

$$Q_r := (b \cdot h) \cdot \left(\frac{H}{2} - \frac{h}{2}\right)$$

$$Q_r = 1.58 \times 10^6 \circ \text{mm}^3$$

Applied rolling shear stress, τ_a

$$\tau_a := \frac{F_v \cdot Q_r}{2 \cdot h \cdot I_{xx}}$$

$$\tau_a = 0.1 \circ N \cdot mm^{-2}$$

$$\tau_{r.g} = 1.23 \circ N \cdot mm^{-2}$$

BS 5268 : Part 2, Clause 30
Permissible panel shear

$$K_{37} := 0.5$$

$$\tau_{adm} := \tau_{r.g} \cdot K_8 \cdot K_{36s} \cdot K_{37}$$

$$\tau_{adm} = 0.61 \circ N \cdot mm^{-2}$$

Rolling shear satisfactory

9. Lateral stability

Clause 2.10.10

Check I_{xx}/I_{yy} ratio

$$\frac{I_{xx}}{I_{yy}} = 20.44$$

Condition (d) applies, i.e. one edge should be held in line

Therefore the section is adequate

Example 7.2 Design of a ply-webbed I-beam

An office building has a flat roof which comprises ply-webbed I-beams spaced at 1.1 m centres with an effective span of 7.6 m. The I-beams support solid-section timber secondary beams that are finished flush with their top and bottom flanges and are positioned perpendicular to their longitudinal axes. Carry out design calculations to show that beams of glued construction with cross-sectional details indicated below in service class 2 are adequate.

Design data
Dead load including
self-weight $= 0.81$ kN/m^2
Imposed load
(medium-term) $= 0.75$ kN/m^2
Flanges: Timber in strength
class C18
Web: Finnish conifer, 9-plies,
12 mm thickness 1.4 mm
veneer sanded with face ply
grain perpendicular to span

Given
$b = 47$ mm
$h = 97$ mm
$t = 12$ mm
$H = 580$ mm

Definitions

Force, kN
Length, m
Cross-sectional dimensions, mm
Stress, Nmm^{-2}

$N := $ newton
$kN := 10^3 \cdot N$
Direction parallel to grain, //
Direction perpendicular to grain, *pp*

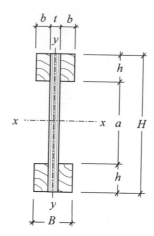

Fig. 7.7 I ply-webbed beam (Example 7.2).

1. *Loading*

Effective length, L_e	$L_e := 7.6 \cdot m$
Beam spacing, Bs	$Bs := 1.1 \cdot m$
Dead load (kN/m²), DL	$DL := 0.81 \cdot kN \cdot m^{-2}$
Imposed load (kN/m²), IL	$IL := 0.75 \cdot kN \cdot m^{-2}$
Total (kN/m²), TL	$TL := DL + IL$
	$TL = 1.56 \circ kN \cdot m^{-2}$
Applied uniformly distributed load (kN/m), w	$w := TL \cdot Bs$
	$w = 1.72 \circ kN \cdot m^{-1}$

2. *K-factors*

Service class 2 (K_2, Table 13)	$K_2 := 1$
Medium-term loading (K_3, Table 14)	$K_3 := 1.25$
No load sharing (K_8, Clause 2.9)	$K_8 := 1$
Width factor, tension member (K_{14}, Clause 2.12.2)	$h := 97 \cdot mm$

$$K_{14} := \left(\frac{300 \cdot mm}{h} \right)^{0.11}$$

$$K_{14} = 1.13$$

Clause 2.10.10

Tension // (K_{27}, Table 22) $\quad K_{27} := 1.1$ for two glued timber sections per flange

Compression // (K_{28c}, Table 22) $\quad K_{28c} := 1.14$ for two glued timber sections per flange

Modulus of elasticity for deflection (K_{28E}, Table 22) $\quad K_{28E} := 1.24$ for a total of four glued timber sections in both flanges

Clause 4.5

Plywood grade stresses (K_{36s}, Table 33)	$K_{36s} := 1.33$
Plywood moduli (K_{36E}, Table 33)	$K_{36E} := 1.54$
Plywood rolling shear (K_{37}, Clause 4.6)	$K_{37} := 0.5$

3. Grade stresses

BS 5268 : Part 2, Table 7
Strength class, C18 timber

Tension // to grain	$\sigma_{timber.t.g.//} := 3.5 \cdot N \cdot mm^{-2}$
Compression // to grain	$\sigma_{timber.c.g.//} := 7.1 \cdot N \cdot mm^{-2}$
Shear // to grain	$\tau_{timber.g.//} := 0.67 \cdot N \cdot mm^{-2}$
Mean E value	$E_{timber.mean} := 9100 \cdot N \cdot mm^{-2}$
Minimum E value	$E_{timber.min} := 6000 \cdot N \cdot mm^{-2}$

BS 5268 : Part 2, Table 41
Finnish conifer, 9-plies 12 mm
   plywood with face grain
   perpendicular to span

Tension perpendicular to face grain	$\sigma_{ply.t.g.pp} := 7.44 \cdot N \cdot mm^{-2}$
Compression perpendicular to face grain	$\sigma_{ply.c.g.pp} := 9.8 \cdot N \cdot mm^{-2}$
Panel shear // and perpendicular to face grain	$\tau_{ply.g} := 3.74 \cdot N \cdot mm^{-2}$
Rolling shear	$\tau_{r.g} := 0.79 \cdot N \cdot mm^{-2}$
Modulus of elasticity in tension and compression perpendicular to face grain	$E_{ply} := 2950 \cdot N \cdot mm^{-2}$
Modulus of rigidity (shear modulus)	$G_w := 270 \cdot N \cdot mm^{-2}$

4. Geometrical Properties

Transform section to an equivalent of all timber.	
Using E_{mean} for timber	$E_{timber} := E_{timber.mean}$
Plywood thickness	$t_{ply} := 12 \cdot mm$
Transformed web thickness, $t_{ply.tfd}$	$t_{ply.tfd} := t_{ply} \cdot \dfrac{E_{ply}}{E_{timber}}$
	$t_{ply.tfd} = 3.89 \circ mm$

Transformed section:

Cross-sectional dimension:

$b := 47 \cdot mm$

$B := t_{ply.tfd} + b \cdot 2$

$B = 97.89 \circ mm$

$H := 580 \cdot mm$

$h := 97 \cdot mm$

$a := H - 2 \cdot h$

$a = 386 \circ mm$

Second moment of area, I_{xx}

$$I_{xx} := \frac{B \cdot H^3}{12} - \frac{(2 \cdot b) \cdot a^3}{12}$$

$$I_{xx} = 1.14 \times 10^9 \circ mm^4$$

Section modulus, Z_{xx}

$$Z_{xx} := \frac{I_{xx}}{\left(\dfrac{H}{2}\right)}$$

$$Z_{xx} = 3.93 \times 10^6 \circ mm^3$$

Second moment of area, I_{yy}

$$I_{yy} := \frac{H \cdot B^3}{12} - \left[\frac{a \cdot \left(\dfrac{2 \cdot b}{2}\right)^3}{12} + \left(a \cdot \frac{2 \cdot b}{2}\right) \right.$$

$$\left. \cdot \left(\frac{t_{ply.tfd}}{2} + \frac{2 \cdot b}{4}\right)^2 \right] \cdot 2$$

$$I_{yy} = 1.52 \times 10^7 \circ mm^4$$

5. *Bending stress*

The maximum applied bending stress in transformed section should not be greater than the lesser of:

(1) Timber tension // to grain

$\sigma_{timber.t.adm.//} := \sigma_{timber.t.g.//} \cdot K_2 \cdot K_3 \cdot K_8 \cdot K_{14} \cdot K_{27}$

$\sigma_{timber.t.adm.//} = 5.5 \circ N \cdot mm^{-2}$

(2) Timber compression// to grain

$\sigma_{timber.c.adm.//} := \sigma_{timber.c.g.//} \cdot K_2 \cdot K_3 \cdot K_8 \cdot K_{28c}$

$\sigma_{timber.t.adm.//} = 10.12 \circ N \cdot mm^{-2}$

(3) Transformed plywood tension perpendicular to face grain

$\sigma_{tfd.ply.t.adm.pp} := \sigma_{ply.t.g.pp} \cdot \dfrac{E_{ply}}{E_{timber}} \cdot K_8 \cdot K_{36s}$

$\sigma_{tfd.ply.t.adm.pp} = 3.21 \circ N \cdot mm^{-2}$

(4) Transformed plywood compression perpendicular to face grain

$\sigma_{tfd.ply.c.adm.pp} := \sigma_{ply.c.g.pp} \cdot \dfrac{E_{ply}}{E_{timber}} \cdot K_8 \cdot K_{36s}$

$\sigma_{tfd.ply.c.adm.pp} = 4.23 \circ N \cdot mm^{-2}$

Thus, the **lowest** value is taken as the permissible stress in bending

$\sigma_{tfd.ply.t.adm.pp} = 3.21 \circ N \cdot mm^{-2}$

Applied load, w	$w = 1.72 \circ \text{kN} \cdot \text{m}^{-1}$
Applied bending moment, M	$M := \dfrac{w \cdot L_e^2}{8}$
	$M = 12.39 \circ \text{kN} \cdot \text{m}$
Applied bending stress, $\sigma_{m.a.\parallel}$	$\sigma_{m.a.\parallel} := \dfrac{M}{Z_{xx}}$

$\sigma_{m.a.\parallel} = 3.15 \circ \text{N} \cdot \text{mm}^{-2}$
Bending stress satisfactory

6. Deflection

BS 5268 : Part 2, Clause 2.10.10

Modulus of elasticity for deflection	$E := E_{timber.min} \cdot K_{28E}$
	$E = 7.44 \times 10^3 \circ \text{N} \cdot \text{mm}^{-2}$
Deflection due to bending	$\Delta_m := \dfrac{5 \cdot w \cdot L_e^4}{384 \cdot E \cdot I_{xx}}$
	$\Delta_m = 8.78 \circ \text{mm}$
Web area, A_w	$A_w := H \cdot t_{ply}$
	$A_w = 6.96 \times 10^3 \circ \text{mm}^2$
Deflection due to shear	$\Delta_s := \dfrac{M}{G_w \cdot A_w}$
	$\Delta_s = 6.59 \circ \text{mm}$
Total deflection	$\Delta_{total} := \Delta_m + \Delta_s$
	$\Delta_{total} = 15.37 \circ \text{mm}$
Permissible deflection	$\Delta_{adm} := 0.003 \cdot L_e$
	$\Delta_{adm} = 22.8 \circ \text{mm}$

Deflection satisfactory

7. Panel shear

Applied shear force, F_v	$F_v := \dfrac{w \cdot L_e}{2}$
	$F_v = 6.52 \circ \text{kN}$
First moment of area above neutral axis, Q	$Q := (2 \cdot b \cdot h) \cdot \left(\dfrac{H}{2} - \dfrac{h}{2} \right)$
	$\quad + \left(t_{ply.tfd} \cdot \dfrac{H}{2} \cdot \dfrac{H}{4} \right)$
	$Q = 2.37 \times 10^6 \circ \text{mm}^3$
Applied panel shear stress, τ_a	$\tau_a := \dfrac{F_v \cdot Q}{I_{xx} \cdot (2 \cdot t_{ply})}$
	$\tau_a = 0.56 \circ \text{N} \cdot \text{mm}^{-2}$

Permissible panel shear

$$\tau_{adm.\parallel} := \tau_{ply.g} \cdot K_8 \cdot K_{36s}$$
$$\tau_{adm.\parallel} = 4.97 \circ N \cdot mm^{-2}$$
Panel shear satisfactory

8. Rolling shear

Applied shear force, F_v

$$F_v = 6.52 \circ kN$$

First moment of area of upper flange above neutral axis, Q_r

$$Q_r := (2 \cdot b \cdot h) \cdot \left(\frac{H}{2} - \frac{h}{2} \right)$$

$$Q_r = 2.2 \times 10^6 \circ mm^3$$

Applied rolling shear stress, τ_a

$$\tau_a := \frac{F_v \cdot Q_r}{2 \cdot h \cdot I_{xx}}$$

$$\tau_a = 0.06 \circ N \cdot mm^{-2}$$

BS 5268 : Part 2, Clause 30
Permissible panel shear

$$K_{37} := 0.5$$
$$\tau_{adm} := \tau_{r.g} \cdot K_8 \cdot K_{36s} \cdot K_{37}$$
$$\tau_{adm} = 0.53 \circ N \cdot mm^{-2}$$
Rolling shear satisfactory

9. Lateral stability

Clause 2.10.10

Check I_{xx}/I_{yy} ratio

$$\frac{I_{xx}}{I_{yy}} = 75.24$$

Condition (f) applies, i.e. compression flanges should be fully restrained, for example, by direct fixing of roof sheeting

Therefore the section is adequate

Chapter 8
Design of Built-up (Spaced) Columns

8.1 Introduction

Timber columns may be classified as simple or built-up, depending on their method of construction. *Simple columns* are fabricated from sawn timber or from glued laminated sections (glulam) to form one piece, for example, a rectangular section. The design requirements for simple columns are discussed in detail in Chapters 5 and 6 for sawn solid sections and glulam members respectively.

Built up columns are those composed of two or more sections of timber connected together either by gluing or by means of mechanical fasteners (such as nails, bolts, etc.) Such columns can provide considerably more strength than the sum of the strength of the sections acting alone. Built-up columns can be constructed in a great variety of shapes, see Fig. 8.1, to meet special needs or to provide larger cross-sections than are ordinarily available, or purely for architectural applications. While glued built-up columns can be assumed to behave as a single member, the full composite strength development in mechanically fastened columns is doubtful. In general mechanically fastened columns may not possess adequate composite action by the connectors to make them act as a unit. In such cases designers are recommended to revert to prototype testing or conservative design solution.[1] BS 5268 : Part 2 : 1996 provides guidance for design of a special type of built-up column known as a *spaced column* which is detailed in this chapter.

Fig. 8.1 Typical cross-sections for built-up elements.

8.2 Spaced columns

Spaced columns are those built up as two or more individual members (shafts) with their longitudinal axes parallel, and are separated at the ends and mid-points with *spacer blocks* and joined together by gluing, nailing, bolting or other means of timber connectors. A typical make-up of a spaced column is illustrated in Fig. 8.2. The spacer blocks are provided to restrain differential movement between the shafts and to maintain their initial spacing under the action of load. The end connectors provide partial fixity to the individual members of the column, thus effectively restraining their buckling tendency. Spaced columns are often used in architectural applications, in trusses as compression chords, and in frame construction.

A spaced column, due to the geometry of its cross-section, could have at least 25% more load-carrying capacity than a single solid member with a similar volume of timber. Spaced columns, apart from being economical, can provide suitable construction through which other pieces such as bracing or column to truss connections can be inserted conveniently.

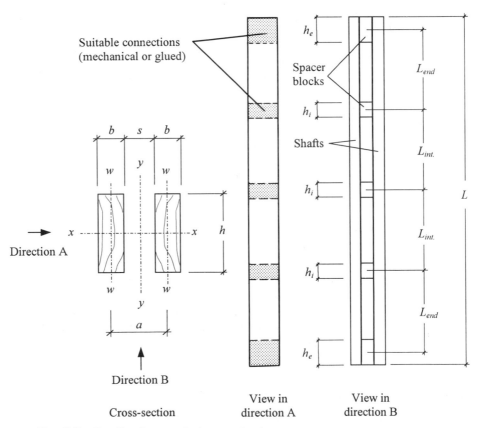

Fig. 8.2 Details of a two-shaft spaced column.

8.3   Design considerations

Design recommendations for spaced columns are detailed in Clauses 2.11.8 to 2.11.11 of BS 5268 : Part 2 : 1996. A summary of the recommendations, which may fall under the following headings, is given below:

(1) geometrical requirements,
(2) modes of failure and permissible loads, and
(3) shear capacity of spacer blocks.

8.3.1   *Geometrical requirements*

With reference to Fig. 8.2, the following criteria must be satisfied:

(1) The clear space between the individual shafts, s, should not be greater than three times the thickness of a single shaft, b (i.e. ensure that $s \leq 3 \times b$).
(2) The length of end spacer blocks, h_e, should be at least six times the thickness of individual shafts (i.e. ensure that $h_e \geq 6 \times b$).
(3) The length of intermediate spacer blocks, h_i, should be not less than 230 mm (i.e. ensure that $h_i \geq 230$ mm).
(4) A sufficient number of spacer blocks should be provided to ensure that the greater slenderness ratio of each shaft between the spacer blocks (L_{end}/i_{ww} and $L_{int.}/i_{ww}$) is limited to 70, or 0.7 times the slenderness ratio of the whole column, whichever is the lesser. Therefore it is necessary to ensure that:

the greater of $\dfrac{L_{end}}{i_{ww}}$ *and* $\dfrac{L_{int.}}{i_{ww}}$ $\leq$ the lesser of 70 *and*

the greater of $0.7\,\dfrac{L_e}{i_{yy}}$ *and* $0.7\,\dfrac{L_e}{i_{xx}}$

where:
L = the overall length of the spaced column
L_e = the effective length of the spaced column
L_{end} = the distance between the centroids of end and closest intermediate spacer blocks
$L_{int.}$ = the distance between the centroids of intermediate spacer blocks
i = radius of gyration with respect to indicated axes, (x–x, y–y, or w–w) – see Fig. 8.2.

8.3.2   *Modes of failure and permissible loads*

Since the following modes of failure are possible, they should be considered and reflected in calculation of the permissible stresses:

Table 8.1 Modification factor K_{13} for the effective length of spaced columns (Table 20, BS 5268 : Part 2)

Method of connection	Value of K_{13}			
	Ratio of space to thickness of the thinner member			
	0	1	2	3
Nailed	1.8	2.6	3.1	3.5
Screwed or bolted	1.7	2.4	2.8	3.1
Connectored	1.4	1.8	2.2	2.4
Glued	1.1	1.1	1.3	1.4

(1) Buckling of the whole column about its x–x axis. In this case the effective length will depend upon full column length and the spacing of the shafts will not improve the performance, hence:

$$L_{e,x-x} = L \times (\text{effective length coefficient, see Table 5.1})$$

The permissible axial and/or bending stresses should be calculated for the equivalent solid column section.

(2) Buckling of the whole column about its y–y axis. In this case the type of connection will affect the magnitude of effective length of the column. Table 20 of BS 5268 : Part 2 : 1996 (reproduced here as Table 8.1) gives the modification factor K_{13} which should be used to calculate the effective length about the y–y axis, hence:

$$L_{e,y-y} = L \times K_{13} \times (\text{effective length coefficient, see Table 5.1})$$

(3) Buckling of an individual shaft about its own w–w axis. For this purpose, the proposed axial load is shared equally between shafts and the effective length is taken to be the larger distance between the centroids of the connections (spacer blocks), i.e. L_{end} or $L_{int.}$, whichever is greater, hence:

$$L_{e,w-w} = \text{the greater of } L_{end} \text{ or } L_{int.}$$

These modes of failure are reflected in permissible load requirements given in the code. It is to be noted that the greatest slenderness ratio (i.e. the greater of L_e/i_{xx}, L_e/i_{yy}, L_{end}/i_{ww} and $L_{int.}/i_{ww}$) will govern the critical failure mode. For example, if L_e/i_{xx} has the largest value, then buckling about the x–x axis is the most critical one and will dictate the magnitude of the permissible stresses.

As before, see Section 5.2, the applied and permissible stresses may be calculated from:

applied compressive stress $\qquad \sigma_{c,a,\parallel} = \dfrac{P}{A}$

permissible compressive stress $\quad \sigma_{c,adm,\parallel} = \sigma_{c,g,\parallel} \times K_2 K_3 K_8 K_{12}$

applied bending stress $\qquad\qquad \sigma_{m,a,\parallel} = \dfrac{M}{Z}$

permissible bending stress $\qquad \sigma_{m,adm,\parallel} = \sigma_{m,g,\parallel} \times K_2 K_3 K_6 K_7 K_8$

8.3.3 *Shear capacity of spacer blocks*

The longitudinal length of the mechanically and/or glue-connected end spacer blocks should not be less than six times the thickness of the individual shaft and should be sufficient to transmit a shear force, F_v, at the interface between each block and one shaft equal to:

$$F_v = \frac{1.3Ab\sigma_{c,a,\parallel}}{na}$$

where:

$A =$ total cross-sectional area of the column
$b =$ thickness of the shaft
$\sigma_{c,a,\parallel} =$ applied compressive stress
$n =$ number of shafts, and
$a =$ distance between centres of adjacent shafts

Therefore applied shear stress, $\tau_{a,\parallel}$, is:

$$\tau_{a,\parallel} = \frac{F_v}{h \cdot h_e}$$

where:

$h =$ width of the shaft, and
$h_e =$ length of the end spacer block

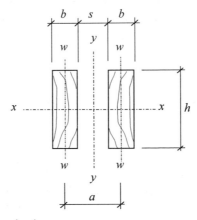

Fig. 8.3 Axes in spaced columns.

and the permissible shear stress, $\tau_{adm,/\!/}$, is:

$$\tau_{adm,/\!/} = \tau_{g,/\!/} \times K_2 K_3 K_8$$

The code also recommends that intermediate spacer blocks should be not less than 230 mm long and be able to transmit a shear force, $F_{v.int.}$, between their individual interfaces equal to half that of the end blocks, i.e. $F_{v.int.} = \frac{1}{2} F_v$.

8.4   Compression members in triangulated frameworks

In triangulated frameworks such as trusses and girders (but excluding trussed rafters designed to Part 3 of BS 5268), Clause 2.11.11 of BS 5268 : Part 2 : 1996 provides guidance as to effective lengths for the compression members and, where these members are considered to be spaced columns, all design should be in accordance with the design requirements for spaced columns except for the design of end connecting joints. In the case of these joints, the shear resistance requirements of Clauses 2.11.9.1.1 and 2.11.9.2 do not apply. The requirements for intermediate spacer blocks for such members are that they should be not less than 200 mm long and should be fixed in such a manner as to transmit a tensile force parallel to the x–x axis, between the individual members, of not less than 2.5% of the total axial force in the spaced compression member.

8.5   Reference

1.   Baird and Ozelton (1984) *Timber Designer's Manual*, 2nd edn. Blackwell Science, Oxford.

8.6   Design examples

Example 8.1   Design of a spaced column subjected to an axial load

A glue-connected spaced column is fabricated from two shafts of equal cross-section of 60 mm × 194 mm, as shown in Fig. 8.4, and carries a medium-term axial compression load of 95 kN. The column is free standing and is unbraced throughout its length against the tendency to buckle in either direction. Check the adequacy of the column if it comprises timber sections in strength class C22 under service class 1 conditions.

Definitions

Force, kN	$N := newton$
Length, m	$kN := 10^3 \cdot N$
Cross-sectional dimensions, mm	Direction parallel to grain, $/\!/$
Stress, Nmm^{-2}	Direction perpendicular to grain, *pp*

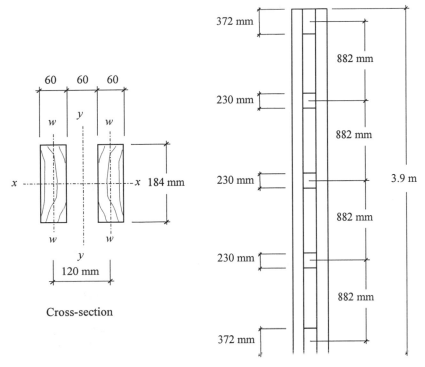

Cross-section

Fig. 8.4 Spaced column details (Example 8.1).

A. *Geometrical requirements*

1. *Geometrical properties*

Column length, L	$L := 3.9 \cdot m$
Number of shafts, n	$n := 2$
Breadth of each shaft, b	$b := 60 \cdot mm$
Width of each shaft, h	$h := 184 \cdot mm$
Space between shafts, s	$s := 60 \cdot mm$

Distance between centres of shafts, a

$$a := s + 2 \cdot \left(\frac{b}{2} \right)$$

$$a = 120 \circ mm$$

Length of end block, h_e	$h_e := 372 \cdot mm$
Length of intermediate block, h_i	$h_i := 230 \cdot mm$
Cross-sectional area of each shaft, A_{shaft}	$A_{shaft} := b \cdot h$
	$A_{shaft} = 1.1 \times 10^4 \circ mm^2$
Total cross-sectional area, A_{total}	$A_{total} := 2 \cdot A_{shaft}$
	$A_{total} = 2.21 \times 10^4 \circ mm^2$
Second moment of area, I_{xx}	$I_{xx} := (2) \cdot \frac{1}{12} \cdot b \cdot h^3$
	$I_{xx} = 6.23 \times 10^7 \circ mm^4$

Second moment of area, I_{yy}	$I_{yy} := \frac{1}{12} \cdot h \cdot (2 \cdot b + s)^3 - \frac{1}{12} \cdot h \cdot s^3$
	$I_{yy} = 8.61 \times 10^7 \circ \text{mm}^4$
For each shaft, I_{ww}	$I_{ww} := \frac{1}{12} \cdot h \cdot b^3$
	$I_{ww} = 3.31 \times 10^6 \circ \text{mm}^4$
Radius of gyration, i_{xx}	$i_{xx} := \sqrt{\dfrac{I_{xx}}{A_{total}}}$
	$i_{xx} = 53.12 \circ \text{mm}$
Radius of gyration, i_{yy}	$i_{yy} := \sqrt{\dfrac{I_{yy}}{A_{total}}}$
	$i_{yy} = 62.45 \circ \text{mm}$
Radius of gyration, i_{ww}	$i_{ww} := \sqrt{\dfrac{I_{ww}}{A_{shaft}}}$
	$i_{ww} = 17.32 \circ \text{mm}$
Effective length of individual shafts about their own w–w axis is the greater of L_{end} and $L_{int.}$	$L_{end} := 882 \cdot \text{mm}$ $L_{int.} := 882 \cdot \text{mm}$
Slenderness ratio about w–w axis, λ_{ww}	$\lambda_{ww} := \dfrac{L_{end}}{i_{ww}}$
	$\lambda_{ww} = 50.92$
Effective length of column about x–x axis, $L_{e.xx}$	$L_{e.xx} := 1.0 \cdot L$ $L_{e.xx} = 3.9 \circ \text{m}$
Slenderness ratio about x–x axis	$\lambda_{xx} := \dfrac{L_{e.xx}}{i_{xx}}$
	$\lambda_{xx} = 73.42$
Effective length of column about y–y axis, $L_{e.yy}$, depends on the shaft/block connections	
BS 5268 : Part 2, Table 20 For glued connections and for s/b ratio of 1	$K_{13} := 1.1$
Effective length of column about y–y axis, $L_{e.yy}$	$L_{e.yy} := 1.0 \cdot L \cdot K_{13}$ $L_{e.yy} = 4.29 \circ \text{m}$
	$\lambda_{yy} := \dfrac{L_{e.yy}}{i_{yy}}$
	$\lambda_{yy} = 68.69$

Condition (1)   Check: $s \le 3 \times b$ $s = 60 \circ \text{mm}$ is less than

$\qquad\qquad\qquad\qquad\qquad\qquad\qquad\qquad 3 \cdot b = 180 \circ \text{mm}$ okay

Condition (2)   Check: $h_e \ge 6 \times b$ $h_e = 372 \circ \text{mm}$ is greater than

$\qquad\qquad\qquad\qquad\qquad\qquad\qquad\qquad 6 \cdot b = 360 \circ \text{mm}$ okay

Condition (3)   Check: $h_i = 230 \circ \text{mm}$ okay
    $h_i \ge 230\,\text{mm}$

Condition (4)   Check slenderness $\lambda_{ww} = 50.92$
ratios so that the greatest $0.7 \cdot \lambda_{xx} = 51.4$
$\lambda_{ww} \leq$ the lesser of (70 and the $0.7 \cdot \lambda_{yy} = 48.09$ Satisfactory
greater of $0.7\lambda_{xx}$ and $0.7\lambda_{yy}$)

B. *Modes of failure and permissible load checks*

From the above calculations it is evident that the greatest slenderness ratio, i.e. $\lambda_{xx} = 69.64$, is the critical one. Therefore buckling about the x–x axis should be considered in determination of the permissible compressive stress.

2. *Grade stresses*

BS 5268 : Part 2, Table 7
Strength class, C22
Compression parallel to grain $\sigma_{c.g.\text{//}} := 7.5 \cdot \text{N} \cdot \text{mm}^{-2}$
Shear parallel to grain $\tau_{g.\text{//}} := 0.71 \cdot \text{N} \cdot \text{mm}^{-2}$
Minimum modulus of $E_{min} := 6500 \cdot \text{N} \cdot \text{mm}^{-2}$
  elasticity, E_{min}

3. *Loading and applied compressive stress*

Medium-term axial loading $P := 95 \cdot \text{kN}$

Applied compressive stress $\sigma_{c.a.\text{//}} := \dfrac{P}{A_{total}}$

$$\sigma_{c.a.\text{//}} = 4.3 \circ \text{N} \cdot \text{mm}^{-2}$$

4. *K-factors*

Service class 1 (K_2, Table 13) $K_2 := 1$
Medium-term loading $K_3 := 1.25$
  (K_3, Table 14)
No load sharing (K_8, Clause 2.9) $K_8 := 1$

Modification factor for the compression member, K_{12}, can be calculated using either one of the following methods:

(1)   Using equation in Annex B:

$$\lambda_{xx} = 73.42 \qquad \sigma_c := \sigma_{c.g.\text{//}} \cdot K_2 \cdot K_3 \qquad \eta := 0.005 \cdot \lambda_{xx} \qquad E := E_{min}$$

$$K_{12.xx} := \frac{1}{2} + \frac{(1+\eta) \cdot \pi^2 \cdot E}{3 \cdot \lambda_{xx}^2 \cdot \sigma_c} - \sqrt{\left(\frac{1}{2} + \frac{(1+\eta) \cdot \pi^2 \cdot E}{3 \cdot \lambda_{xx}^2 \cdot \sigma_c} \right)^2 - \frac{\pi^2 \cdot E}{1.5 \cdot \lambda_{xx}^2 \cdot \sigma_c}}$$

$$K_{12.xx} = 0.52$$

(2)   Alternatively, using Table 19:

$$\frac{E_{min}}{\sigma_{c.g.\parallel} \cdot K_2 \cdot K_3} = 693.33 \quad \text{and} \quad \lambda_{xx} = 73.42 \qquad\qquad K_{12.xx} = 0.52$$

Permissible compressive
stress

$\sigma_{c.adm.\parallel.xx} := \sigma_{c.g.\parallel} \cdot K_2 \cdot K_3 \cdot K_8 \cdot K_{12.xx}$
$\sigma_{c.adm.\parallel.xx} = 4.83 \circ N \cdot mm^{-2}$
Satisfactory

C.   *Check shear capacity of spacer blocks*

BS 5268 : Part 2, Clause 2.11.9

Shear force acting at end
blocks, F_v

$$F_v := \frac{1.3 \cdot A_{total} \cdot b \cdot \sigma_{c.a.\parallel}}{n \cdot a}$$

$F_v = 30.88 \circ kN$

Shear force for intermediate
blocks, $F_{v.int.}$

$$F_{v.int.} := \frac{F_v}{2}$$

$F_{v.int.} = 15.44 \circ kN$

Shear area for end
blocks $= h_e \times h$
Shear area for intermediate
blocks $= h_i \times h$

$A_{s.end} := h_e \cdot h$
$A_{s.end} = 6.84 \times 10^4 \circ mm^2$
$A_{s.int.} := h_i \cdot h$
$A_{s.int.} = 4.23 \times 10^4 \circ mm^2$

Shear stress at end
blocks, $\tau_{a,\parallel,end}$

$$\tau_{a.\parallel.end} := \frac{F_v}{A_{s.end}}$$

$\tau_{a.\parallel.end} = 0.45 \circ N \cdot mm^{-2}$

Shear stress at intermediate
blocks, $\tau_{a,\parallel,int}$

$$\tau_{a.\parallel.int.} := \frac{F_{v.int.}}{A_{s.int.}}$$

$\tau_{a.\parallel.int.} = 0.36 \circ N \cdot mm^{-2}$

Permissible shear stress, $\tau_{adm,\parallel}$

$\tau_{adm} := \tau_{g.\parallel} \cdot K_2 \cdot K_3 \cdot K_8$
$\tau_{adm} = 0.89 \circ N \cdot mm^{-2}$
Shear satisfactory

The spaced column is adequate

Example 8.2   *Design of a spaced column, determination of axial load capacity and effect of lateral loading*

A nail connected spaced column is fabricated from two shafts of equal cross-section of 63 mm × 225 mm, as shown in Fig. 8.5. Assuming that the column is a free standing one and is unbraced throughout its length against the tendency to buckle in either direction,
(a)   check the adequacy of the column if it comprises timber sections in strength class C24 under service class 2 conditions,

(b)   determine the maximum axial load, under long-duration loading, that the column can carry, and

(c)   check that the section is adequate to resist a very short-term lateral wind load of 1.25 kN/m of height tending to bend the column about its x–x axis together with an axial load of 45 kN.

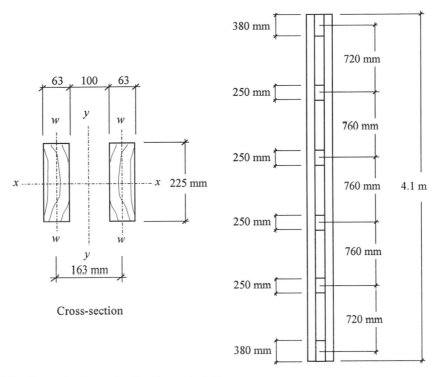

Fig. 8.5   Spaced column details (Example 8.2).

Definitions

Force, kN	$N := $ newton
Length, m	$kN := 10^3 \cdot N$
Cross-sectional dimensions, mm	Direction parallel to grain, $//$
Stress, Nmm^{-2}	Direction perpendicular to grain, pp

A.   *Geometrical requirements*

1.   *Geometrical properties*

Column length, L	$L := 4.1 \cdot m$
Number of shafts, n	$n := 2$
Breadth of each shaft, b	$b := 63 \cdot mm$
Width of each shaft, h	$h := 225 \cdot mm$
Space between shafts, s	$s := 100 \cdot mm$

Distance between centres of
shafts, a

$$a := s + 2 \cdot \left(\frac{b}{2} \right)$$

$a = 163 \circ \text{mm}$

Length of end block, h_e $h_e := 380 \cdot \text{mm}$

Length of intermediate block, h_i $h_i := 250 \cdot \text{mm}$

Cross-sectional area of each
shaft, A_{shaft}

$A_{shaft} := b \cdot h$

$A_{shaft} = 1.42 \times 10^4 \circ \text{mm}^2$

Total cross-sectional area, A_{total}

$A_{total} := 2 \cdot A_{shaft}$

$A_{total} = 2.84 \times 10^4 \circ \text{mm}^2$

Second moment of area, I_{xx}

$I_{xx} := (2) \cdot \frac{1}{12} \cdot b \cdot h^3$

$I_{xx} = 1.2 \times 10^8 \circ \text{mm}^4$

Second moment of area, I_{yy}

$I_{yy} := \frac{1}{12} \cdot h \cdot (2 \cdot b + s)^3 - \frac{1}{12} \cdot h \cdot s^3$

$I_{yy} = 1.98 \times 10^8 \circ \text{mm}^4$

For each shaft, I_{ww}

$I_{ww} := \frac{1}{12} \cdot h \cdot b^3$

$I_{ww} = 4.69 \times 10^6 \circ \text{mm}^4$

Radius of gyration, i_{xx}

$$i_{xx} := \sqrt{\frac{I_{xx}}{A_{total}}}$$

$i_{xx} = 64.95 \circ \text{mm}$

Radius of gyration, i_{yy}

$$i_{yy} := \sqrt{\frac{I_{yy}}{A_{total}}}$$

$i_{yy} = 83.5 \circ \text{mm}$

Radius of gyration, i_{ww}

$$i_{ww} := \sqrt{\frac{I_{ww}}{A_{shaft}}}$$

$i_{ww} = 18.19 \circ \text{mm}$

Effective length of individual
shafts about their own minor
w–w axis is the greater of
L_{end} and $L_{int.}$

$L_{end} := 720 \cdot \text{mm}$

$L_{int.} := 760 \cdot \text{mm}$

Slenderness ratio about w–w
axis, λ_{ww}

$$\lambda_{ww} := \frac{L_{int.}}{i_{ww}}$$

$\lambda_{ww} = 41.79$

Effective length of column
about x–x axis, $L_{e.xx}$

$L_{e.xx} := 1.0 \cdot L$

$L_{e.xx} = 4.1 \circ \text{m}$

Slenderness ratio about x–x axis

$$\lambda_{xx} := \frac{L_{e.xx}}{i_{xx}}$$

$\lambda_{xx} = 63.12$

Effective length of column about
y–y axis, $L_{e.yy}$, depends on the
shaft/block connections

BS 5268 : Part 2, Table 20

For nailed connections and for
s/b ratio of 1.59

$K_{13} := 2.895$

Effective length of column about $y–y$ axis, $L_{e.yy}$

$L_{e.yy} := 1.0 \cdot L \cdot K_{13}$
$L_{e.yy} = 11.87 \circ m$

Slenderness ratio about $y–y$ axis:

$\lambda_{yy} := \dfrac{L_{e.yy}}{i_{yy}}$

$\lambda_{yy} = 142.14$

Condition (1)   Check: $s \leq 3 \times b$ $s = 100 \circ mm$ is less than

$3 \cdot b = 189 \circ mm$ okay

Condition (2)   Check: $h_e \geq 6 \times b$ $h_e = 380 \circ mm$ is greater than

$6 \cdot b = 378 \circ mm$ okay

Condition (3)   Check: $h_i \geq 230\,mm$ $h_i = 250 \circ mm$ okay

Condition (4)   Check slenderness
ratios so that the greatest
$\lambda_{ww} \leq$ the lesser of (70 and the
greater of $0.7\lambda_{xx}$ and $0.7\lambda_{yy}$)

$\lambda_{ww} = 41.79$
$0.7 \cdot \lambda_{xx} = 44.19$
$0.7 \cdot \lambda_{yy} = 99.5$ Satisfactory

B. *Modes of failure and permissible load checks*

From the above calculations it is evident that the greatest slenderness ratio, i.e. $\lambda_{yy} = 142.14$, is the critical one. Therefore buckling about the $y–y$ axis should be considered in determination of the permissible compressive stress.

2. *Grade stresses*

BS 5268 : Part 2, Table 7
Strength class, C24
Compression parallel to grain
Shear parallel to grain
Bending parallel to grain
Minimum modulus of
elasticity, E_{min}

$\sigma_{c.g.//} := 7.9 \cdot N \cdot mm^{-2}$
$\tau_{g.//} := 0.71 \cdot N \cdot mm^{-2}$
$\sigma_{m.g.//} := 7.5 \cdot N \cdot mm^{-2}$
$E_{min} := 7200 \cdot N \cdot mm^{-2}$

3. *K-factors*

Service class 2 (K_2, Table 13)
Long-term loading
  (K_3, Table 14)
Very short-term loading
  (K_3, Table 14)
No load sharing (K_8, Clause 2.9)

$K_2 := 1$
$K_{3.long} := 1$

$K_{3.v.short} := 1.75$

$K_8 := 1$

Modification factor for the compression member, K_{12}, can be calculated using either one of the following methods:

(1) Using equation in Annex B:

$$\lambda_{yy} = 142.14 \qquad \sigma_c := \sigma_{c.g.//} \cdot K_2 \cdot K_{3.long} \qquad \eta := 0.005 \cdot \lambda_{yy} \qquad E := E_{min}$$

$$K_{12.yy} := \frac{1}{2} + \frac{(1+\eta) \cdot \pi^2 \cdot E}{3 \cdot \lambda_{yy}^2 \cdot \sigma_c} - \sqrt{\left(\frac{1}{2} + \frac{(1+\eta) \cdot \pi^2 \cdot E}{3 \cdot \lambda_{yy}^2 \cdot \sigma_c}\right)^2 - \frac{\pi^2 \cdot E}{1.5 \cdot \lambda_{yy}^2 \cdot \sigma_c}}$$

$$K_{12.yy} = 0.23$$

(2) Alternatively, using Table 19:

$$\frac{E_{min}}{\sigma_{c.g.//} \cdot K_2 \cdot K_{3.long}} = 911.39 \quad \text{and} \quad \lambda_{yy} = 142.14 \qquad\qquad K_{12.yy} = 0.23$$

Permissible compressive stress	$\sigma_{c.adm.//} := \sigma_{c.g.//} \cdot K_2 \cdot K_{3.long} \cdot K_8 \cdot K_{12.yy}$ $\sigma_{c.adm.//} = 1.84 \circ N \cdot mm^{-2}$
Allowable axial compressive load, P_{adm}	$P_{adm} := \sigma_{c.adm.//} \cdot A_{total}$ $P_{adm} = 52.14 \circ kN$
Applied compressive stress	$\sigma_{c.a.//} := \sigma_{c.adm.//}$ $\sigma_{c.a.//} = 1.84 \circ N \cdot mm^{-2}$

C. *Check shear capacity of spacer blocks*

BS 5268 : Part 2, Clause 2.11.9

Shear force acting at end blocks, F_v	$F_v := \dfrac{1.3 \cdot A_{total} \cdot b \cdot \sigma_{c.a.//}}{n \cdot a}$ $F_v = 13.1 \circ kN$
Shear force for intermediate blocks, $F_{v.int.}$	$F_{v.int.} := \dfrac{F_v}{2}$ $F_{v.int.} = 13.72 \circ kN$
Shear area for end blocks	$A_{s.end} := h_e \cdot h$ $A_{s.end} = 8.55 \times 10^4 \circ mm^2$
Shear area for intermediate blocks	$A_{s.int.} := h_i \cdot h$ $A_{s.int.} = 5.63 \times 10^4 \circ mm^2$
Shear stress at end blocks, $\tau_{a,//,end}$	$\tau_{a.//.end} := \dfrac{F_v}{A_{s.end}}$ $\tau_{a.//.end} = 0.15 \circ N \cdot mm^{-2}$
Shear stress at intermediate blocks, $\tau_{a,//,int.}$	$\tau_{a.//.int.} := \dfrac{F_{v.int.}}{A_{s.int.}}$ $\tau_{a.//.int.} = 0.12 \circ N \cdot mm^{-2}$
Permissible shear stress, $\tau_{adm,//}$	$\tau_{adm} := \tau_{g.//} \cdot K_2 \cdot K_{3.long} \cdot K_8$ $\tau_{adm} = 0.71 \circ N \cdot mm^{-2}$ Shear satisfactory

The axial load-carrying capacity of the spaced column is $P_{adm} = 52.14 \circ kN$

In part (c) of Example 8.2 it is required to check the effects of axial load and bending moment due to lateral loading about x–x axis:

4. Prescribed loading

Axial load, P	$P := 45 \cdot kN$
Very short-term lateral uniformly distributed load, w	$w := 1.25 \cdot kN \cdot m^{-1}$

5. Compression and bending stresses

Applied compressive stress

$$\sigma_{c.a.//} := \frac{P}{A_{total}}$$

$$\sigma_{c.a.//} = 1.59 \circ N \cdot mm^{-2}$$

Section modulus, Z_{xx}

$$Z_{xx} := \frac{2 \cdot b \cdot h^2}{6}$$

$$Z_{xx} = 1.06 \times 10^6 \circ mm^3$$

Applied bending moment

$$M := \frac{w \cdot L^2}{8}$$

$$M = 2.63 \circ kN \cdot m$$

Applied bending stress

$$\sigma_{m.a.//} := \frac{M}{Z_{xx}}$$

$$\sigma_{m.a.//} = 2.47 \circ N \cdot mm^{-2}$$

Modification factor for compression member, K_{12}, for a very short-duration loading considering the largest slenderness ratio, i.e. λ_{yy}, can be calculated using either one of the following methods:

(1) Using equation in Annex B:

$$\lambda_{yy} = 142.14 \qquad \sigma_c := \sigma_{c.g.//} \cdot K_2 \cdot K_{3.v.short} \qquad \eta := 0.005 \cdot \lambda_{yy} \qquad E := E_{min}$$

$$K_{12.yy} := \frac{1}{2} + \frac{(1+\eta) \cdot \pi^2}{3 \cdot \lambda_{yy}^2} \cdot \frac{E}{\sigma_c} - \sqrt{\left(\frac{1}{2} + \frac{(1+\eta) \cdot \pi^2}{3 \cdot \lambda_{yy}^2} \cdot \frac{E}{\sigma_c} \right)^2 - \frac{\pi^2}{1.5 \cdot \lambda_{yy}^2} \cdot \frac{E}{\sigma_c}}$$

$$K_{12.yy} = 0.15$$

(2) Alternatively, using Table 19:

$$\frac{E_{min}}{\sigma_{c.g.//} \cdot K_2 \cdot K_{3.v.short}} = 520.8 \quad \text{and} \quad \lambda_{yy} = 142.14 \qquad K_{12.yy} = 0.15$$

Check interaction quantity:

Permissible compressive stress

$$\sigma_{c.adm.//.yy} := \sigma_{c.g.//} \cdot K_2 \cdot K_{3.v.short} \cdot K_8 \cdot K_{12.yy}$$

$$\sigma_{c.adm.//.yy} = 2.05 \circ N \cdot mm^{-2}$$

Permissible bending stress

$$\sigma_{m.adm.//} := \sigma_{m.g.//} \cdot K_2 \cdot K_{3.v.short} \cdot K_8$$

$$\sigma_{m.adm.//} = 13.13 \circ N \cdot mm^{-2}$$

Euler critical stress

$$\sigma_e := \frac{\pi^2 \cdot E_{min}}{\lambda_{yy}^2}$$

$$\sigma_e = 3.52 \circ N \cdot mm^{-2}$$

Interaction quantity

$$\frac{\sigma_{m.a.\//}}{\sigma_{m.adm.\//} \cdot \left(1 - \frac{1.5 \cdot \sigma_{c.a.\//}}{\sigma_e} \cdot K_{12.yy}\right)} + \frac{\sigma_{c.a.\//}}{\sigma_{c.adm.\//.yy}} = 0.98 < 1 \quad \text{okay}$$

6. *Check deflection*

Deflection due to bending

$$\Delta_m := \frac{5 \cdot w \cdot L_{e.xx}^4}{384 \cdot E_{min} \cdot I_{xx}}$$

$$\Delta_m = 5.34 \circ mm$$

Deflection due to shear

$$\Delta_s := \frac{19.2 \cdot M}{(b \cdot h) \cdot 2 \cdot E_{min}}$$

$$\Delta_s = 0.25 \circ mm$$

Total deflection

$$\Delta_{total} := \Delta_m + \Delta_s$$

$$\Delta_{total} = 5.59 \circ mm$$

Permissible deflection

$$\Delta_{adm} := 0.003 \cdot L_{e.xx}$$

$$\Delta_{adm} = 12.3 \circ mm$$

Deflection satisfactory

The spaced column is adequate

Chapter 9
Design of Timber Connections

9.1 Introduction

In early timber buildings, many of which are still around, individual members were joined together by contact joints and wooden pins or iron dowels. Techniques such as half-lap, cogging, framed, scarf and tenon joinery joints (see Fig. 9.1) utilising metal screws hidden by wooden pegs provided the required connection capacity. The connection schemes were simple primarily because the building systems were simple, with nearly symmetrical arrangements of structural elements. The present method of building construction has since evolved considerably and requires more complex connection systems and, as such, the designer should be aware of the structural behaviour and material characteristics of the connection components.

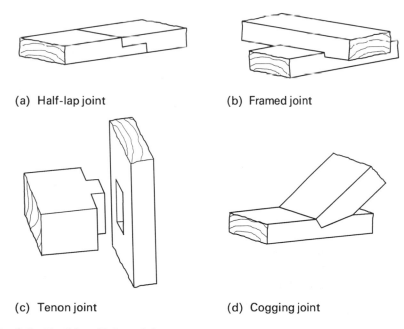

(a) Half-lap joint   (b) Framed joint

(c) Tenon joint   (d) Cogging joint

Fig. 9.1   Traditional joinery joints.

The common connecting systems in structural timber may be classified as mechanically fastened and glued joints. Mechanical fasteners may also be divided into two groups depending on how they transfer the forces between the connected members. These being *dowel-type* fasteners, such as nails, screws, bolts and dowels, and *connectors*, such as toothed-plates, split-rings, shear plates and punched metal plates in which the load transmission is primarily achieved by a large bearing area at the surface of the members.

In this chapter details of the general considerations necessary for the design of dowel-type fasteners are described. This is followed by details of the recommendations for the design of connectored and glued timber joints.

9.2   General design considerations

As mentioned in Chapter 2, the most recent revision of Part 2 of BS 5268 in 1996 was to bring this code as close as possible and to run in parallel with DD ENV 1995-1-1 *Eurocode 5: Design of timber structures. Part 1.1 General rules and rules for buildings* (EC5). The overall aim has been to incorporate material specifications and design approaches from EC5, while maintaining a permissible stress code which designers, accustomed to BS 5268, will feel familiar with and be able to use without difficulty. In this respect design methodology for dowel-type connections has been subjected to considerable change and while it remains a permissible stress design method, it takes the EC5 approach to design of nailed, screwed, bolted and dowelled connections.

Design recommendations for timber joints are detailed in Section 6 of BS 5268: Part 2. This section provides the basic lateral shear and withdrawal loads for the most common dowel type fasteners in Tables 54–74. The values given in these tables have been calculated using the equations and rules detailed in EC5. A summary of EC5 rules and equations is given in Annex G of BS 5268 : Part 2. These equations can be used in lieu of the tabulated data if preferred, and may be used for other material/fastener applications that are not currently detailed in the code, for example, the use of bronze or copper and high tensile steel dowels. In this chapter design calculations are based on the tabulated values; while the use of equations is discussed in Chapter 10 which outlines the Eurocode approach to design of timber structures.

A summary of the general design considerations for timber joints detailed in Section 6 of the code is as follows:

(1)   Joints should be designed so that the loads induced in each fastener or timber connector unit by the design loads appropriate to the structure should not exceed the permissible values determined in accordance with Section 6 of the code.

(2)   When more than one nail, screw, bolt, etc. is used in a joint, the permissible load is the sum of the permissible loads for the individual

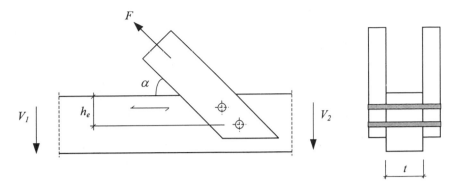

Fig. 9.2   Shear stress in the jointed timber.

units providing that the centroid of the fasteners' group lies on the axis of the member, and that the spacing, edge and end distances are sufficient to develop the full strength of each fixing unit.

(3)   If the line of action of a force in a member does not pass through the centroid of the fasteners' group, account should be taken of the stresses and rotation due to the secondary moments induced by the eccentricity. This will impose higher loads on the fasteners furthest from the centroid of the group.

(4)   If the load on a joint is carried by more than one type of fastener, due account should be taken of the relative stiffnesses. Glue and mechanical fasteners have very different stiffness properties and should not be assumed to act in unison.

(5)   For the purpose of designing joints in glued laminated timber, tabulated values should be taken for the strength class of timber from which the glued laminated timber was made.

(6)   The effective cross-section of a jointed member should be used when calculating its strength. In addition, the shear stress condition shown in Fig. 9.2 should be satisfied in the jointed member. Therefore the design requirement is to ensure that:

$$V \leq \tfrac{2}{3} \cdot \tau_{adm} \cdot h_e \cdot t \qquad (9.1)$$

where:

V = maximum shear force per member at the joint where
$\quad (V_1 + V_2 = F \sin \alpha)$
h_e = distance from the loaded edge to the furthest fastener
τ_{adm} = permissible shear stress, and
t = member thickness.

9.3   Joint slip

All mechanically connected timber joints, under the action of applied load, will undergo slip which occurs between their component parts. Slip can be a

Table 9.1 Fastener slip moduli K_{ser} per fastener per shear plane for service classes 1 and 2 (Table 52, BS 5268 : Part 2)

Fastener type	K_{ser}	
	Timber-to-timber or panel-to-timber (N/mm)	Steel-to-timber[a] (N/mm)
Nails (no pre-drilling)	$\frac{1}{25}\rho^{1.5}d^{0.8}$	$\frac{1}{15}\rho^{1.5}d^{0.8}$
Nails (pre-drilled) Screws Bolts and dowels	$\frac{1}{20}\rho^{1.5}d$	$\frac{1}{12}\rho^{1.5}d$

where ρ is the joint member characteristics density given in Table 7 (in kg/m^3)
$\quad\quad d$ is the fastener diameter (in mm)
[a] For the values in this column to apply, the holes in the steel plate should be the minimum clearance diameter and the steel plate should have a thickness of 1.2 mm or 0.3 times the fastener diameter, whichever is the greater

significant factor in the deformation and, in some cases, it can also affect the strength of some components. Clause 6.2 of BS 5268 : Part 2 provides a method for calculation of slip in joints, whereby the slip per fastener per shear plane, u, in millimetres, may be obtained from:

$$u = \frac{F}{K_{ser}} \tag{9.2}$$

where:
$\quad F = $ applied load per fastener per shear plane, in Newtons (N)
$\quad K_{ser} = $ slip modulus, in Newtons per millimetre per fastener per shear plane, given in Table 52 of the code (reproduced here as Table 9.1).

For bolted joints it is recommended that an additional 1 mm should be added at each joint to allow for take-up of bolt hole tolerances.
It is to be noted that the magnitude of slip is time-dependent and the final slip following medium-term or long-term loading may be considerably greater than the slip calculated as above, particularly in conditions of fluctuating moisture content.

9.4 Effective cross-section

The effective cross-section of a jointed member should be used when calculating its strength. This is because several connector units remove part of the cross-sectional area of the timber and hence affecting the load capacity of the timber section. In such cases the effective cross-section should be determined by deducing the net projected area of the connectors from the gross area of the cross-section being considered.

The recommendations of BS 5268 : Part 2 for multi-dowel-type joints are as follow:

(1) *Nailed and screwed joints.* When assessing the effective cross-section of multi-nailed or multi-screwed joints, all nails or screws of 5 mm in diameter, or larger, that lie within a distance of five nail or screw diameters, measured parallel to the grain from a given cross-section, should be considered as occurring at that cross-section.

No reduction of cross-section needs to be made for screws of less than 5 mm diameter, and for nails of the same diameter but driven without pre-drilling.

(2) *Bolted and dowelled joints.* When assessing the effective cross-section of multi-bolted or multi-dowelled joints, all bolts or dowels that lie within a distance of two bolt or dowel diameters, measured parallel to the grain from a given cross-section, should be considered as occurring at that cross-section.

(3) *Connectored joints.* When assessing the effective cross-section of multi-connectored joints, such as *toothed-plate, split-ring* and *shear-plate connector joints*, all connectors and their bolts that lie within a distance of 0.75 times the nominal size of the connector measured parallel to the grain from a given cross-section should be considered as occurring at that cross-section.

9.5 Spacing rules

In order to avoid splitting of timber, all dowel-type fasteners must be spaced at suitable distances from each other and from the ends and edges of timber or wood-based material. The various distances involved are illustrated in Fig. 9.3.

An end distance is said to be loaded when the load on the fastener has a component towards the end of the timber. Otherwise, it is referred to as an unloaded end distance. Loaded end distances need to be greater than unloaded ones. In a similar way the edge distance may be loaded or unloaded.[1,2]

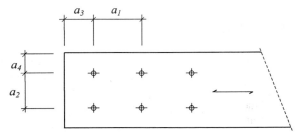

Fig. 9.3 Spacing of dowel-type connectors.

The values of spacings and distances vary from one fastener type to another, as well as between the various material types. The spacings and distances recommended in Section 6 of BS 5268 : Part 2 are detailed below:

(1) *Nail spacing.* The recommended end and edge distances and spacing of nails, to avoid undue splitting, are given in Table 53 of the code (reproduced here as Table 9.2).

Pre-drilling (with a hole diameter up to 0.8 times the nail diameter) reduces the splitting tendency considerably and hence allows much closer spacing of the nails.

All softwoods, except Douglas fir, may have values of spacing (but not edge distances) for timber to timber joints multiplied by 0.8. For nails driven into a glued laminated section at right angles to the glue surface, the spacings (but not edge distances) can be multiplied by a further factor of 0.9.

(2) *Screw spacing.* The recommended end and edge distances and spacing of screws, to avoid undue splitting, are given in Table 59 of the code (reproduced here as Table 9.3).

(3) *Bolt and dowel spacings.* The recommended end and edge distances and spacing of bolts and dowels are given in Table 75 of the code (reproduced here as Table 9.4).

Table 9.2   Minimum nail spacings (Table 53, BS 5268 : Part 2)

Spacing[a]	Timber-to-timber joints		Steel plate-to-timber joints	Wood-based panel products[b] to-timber joints
	Without pre-drilled holes	With pre-drilled holes	Without pre-drilled holes	Without pre-drilled holes
End distance parallel to grain (a_3)	20d	10d	14d	14d
Edge distance perpendicular to grain (a_4)	5d	5d	5d	c
Distance between lines of nails, perpendicular to grain (a_2)	10d	3d	7d	7d
Distance between adjacent nails in any one line, parallel to grain (a_1)	20d	10d	14d	14d

where d is the nail diameter
[a] Spacings and distances are illustrated in Fig. 9.3
[b] Plywood, tempered hardboard or particleboard
[c] The loaded edge distance should not be less than: in timber 5d, in plywood 3d, in particleboard and tempered hardboard 6d, and in all other cases the edge distance should not be less than 3d

Table 9.3 Minimum screw spacings, (Table 59, BS 5268 : Part 2)

Spacing[a]	With pre-drilled holes
End distance parallel to grain (a_3)	$10d$
Edge distance perpendicular to grain (a_4)	$5d$
Distance between lines of screws, perpendicular to grain (a_2)	$3d$
Distance between adjacent screws in any one line, parallel to grain (a_1)	$10d$

where d is the shank diameter of the screw
[a] Spacings and distances are illustrated In Fig. 9.3

Table 9.4 Minimum bolt and dowel spacings (Table 75 and Clause 6.6.7.3, BS 5268 : Part 2)

Direction of loading	End distance		Edge distance		Distance between bolts or rows of bolts	
	Loaded	Unloaded	Loaded	Unloaded	Across the grain	Parallel to grain
Loading // to grain	$7d$	$4d$	$1.5d$	$1.5d$	$4d$	$4d$
Loading ⊥ to grain	$4d$	$4d$	$4d$	$1.5d$	$4d$	$5d$[a]

[a] Where the member thickness, t, is less than 3 times the bolt diameter, d, the spacing may be taken as the greater of $3d$ or $(2 + t/d)d$.

(4) *Connectors spacings*. Associated with each type and size of *toothed-plate*, *split-ring* and *shear-plate connectors* are standard end and edge distances and spacing between connectors which permit the basic load to be used. These standard distances are given in Tables 77–79 for toothed-plate connectors and in Tables 85–87 for both split-ring and shear-plate connectors. If the end distance, edge distance or spacing is less than the standard, then the basic load should be modified. For further details readers are referred to Clauses 6.7–6.9 of BS 5268 : Part 2 : 1996.

9.6 Multiple shear lateral loads

Dowel-type fasteners, such as nails, screws, bolts and dowels, are used to hold two, three or more members together to form a joint. In general, they are designed to carry lateral shear loads, but there are occasions where they might be subjected to axial loads (or withdrawal loads in the case of nailed or screwed joints) or a combination of lateral shear and axial loads.

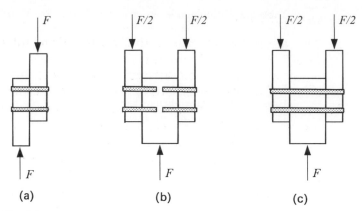

Fig. 9.4   Single shear (a) and (b) and double shear (c) joints.

Dowel-type fasteners, depending on their type and size and the arrangement of the joint components, may be subjected to single, double or multiple lateral shear forces (see Fig. 9.4). It is important to note that the dowels in joints (a) and (b), shown in Fig. 9.4, are all in single shear (i.e. one shear plane per dowel) and the dowels in joint (c) are in double shear (i.e. two shear planes per dowel).

The basic multiple shear lateral load for each dowel can generally be obtained by multiplying the value for its basic single shear load (given in the relevant tables of the code) by the number of shear planes; provided that the recommended standard thickness for members of the joint and the recommended joint details are met.

9.7   Nailed joints

Nails are straight slender fasteners, usually pointed and headed, and are available in a variety of lengths and cross-sectional areas. There are also many types and forms of nails; but the most frequently used nails are the bright, smooth, common steel wire nails with circular cross-sectional area with a minimum tensile strength of $600 \, \text{N/mm}^2$, which are available in standard sizes of 2.65 mm to 8 mm in diameter. Nails can be plain or enamelled, etched, electroplated, galvanised or polymer coated, for example, for protection against corrosion.

Nailed joints, which should normally contain at least two nails, are simple to fabricate and are of considerable economic value. They are suitable for lightly loaded structures and for thin members. They are commonly used in framing, walls, decks, floors and roofs and in nearly every construction that involves light loads and simple elements. For the outer (headside) members larger than 50 mm thick, nails are generally replaced by bolts to reduce splitting of the timber.

9.7.1 *Improved nails*

The performance of a nail, both under lateral and withdrawal loading, may be enhanced by mechanically deforming nail shanks to form *annular ringed shank* or *helical threaded shank* nails (see Fig. 9.5). Such nails provide higher withdrawal resistance than plain shank nails of the same size. Other forms of improved nails are obtained by *grooving* or *twisting* of square cross-sectioned nails. The process of twisting not only modifies the nail surface but also work-hardens the steel, increasing its yield strength.

In order to provide for the enhanced performance of improved nails, BS 5268 : Part 2, Clause 6.4.4.4 permits a 20% increase in the basic single shear lateral load capacities of the square grooved and square twisted shank nails, and a 50% increase in the basic withdrawal load capacities for the threaded part of annular ringed shank and helical threaded shank nails.

9.7.2 *Pre-drilling*

Nails driven into dense timbers are likely to induce timber splitting. Various research studies have shown that pre-drilling can prevent timber splitting and also has other advantages of less slip in the joint, increase in lateral load-carrying capacity and permits a more densely nailed joint (i.e. reduced nail spacings and distances).

The recommendations of BS 5268 : Part 2 for pre-drilling are as follows:

(1) Hardwoods in strength classes of D30 to D70 will usually require pre-drilling and the code recommends that the diameter of pre-drilled holes should not be greater than 0.8 times the nail diameter.

   In general it is usual to provide pre-drilling when the density of the timber is 500 kg/m^3 or greater.

(2) For nails driven into pre-drilled holes in softwood strength classes C14 to C40 and TR26, the code permits a 15% increase in their load capacities.

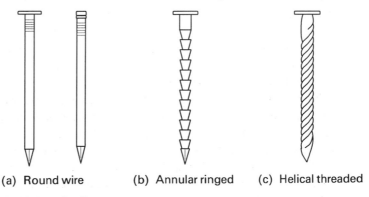

(a) Round wire   (b) Annular ringed   (c) Helical threaded

Fig. 9.5 Types of nails.

Table 9.5 Basic single shear lateral loads for round wire nails in a timber-to-timber joint (Table 54, BS 5268 : Part 2)

Nail diameter (mm)	Softwoods (not pre-drilled)					Hardwoods (pre-drilled)		
	Standard penetration[a] (mm)	Basic single shear lateral load, (N) Strength class				Minimum penetration[a] (mm)	Basic single shear lateral load, (N) Strength class	
		C14	C16/18/22	C24	C27/30/35/40		D30/35/40	D50/60/70
2.65	32	242	251	266	274	21	374	414
3.00	36	297	308	327	336	24	467	517
3.35	40	357	369	392	403	27	569	630
3.75	45	430	444	472	486	30	695	770
4.00	48	478	494	525	540	32	780	864
4.50	54	581	600	638	656	36	961	1065
5.00	60	691	714	759	781	40	1159	1284
5.60	67	833	861	915	941	45	1417	1569
6.00	72	934	965	1026	1054[b]	48	1601	1773
6.70	80	1120	1158	1230	1265[b]	54	1945	2154
8.00	96	1501	1552	1649	1695[b]	64	2658	2944

[a] These values apply to both the headside thickness and pointside penetration
[b] Holes should be pre-drilled

9.7.3 *Basic single shear lateral loads*

1. Timber-to-timber nailed joints

The basic single shear lateral loads for single round wire nails with a minimum tensile strength of $600\,\text{N/mm}^2$, driven at right angles to the side grain of timber in service classes 1 and 2, are given in Table 54 of the code (reproduced here as Table 9.5).

For nails driven into the end grain of timber, the values given in Table 54 should be multiplied by the end grain modification factor K_{43} which has a value of 0.7, i.e. $K_{43} = 0.7$.

For nails driven into pre-drilled holes in softwood strength classes C14 to C40 and TR26, the basic single shear lateral load values given in Table 54 may be multiplied by 1.15, i.e. for convenience introduce a modification factor for pre-drilling, as $K_{pre-drill} = 1.15$.

For square grooved and square twisted shank nails of steel with a yield stress of not less than $375\,\text{N/mm}^2$, the basic single shear lateral loads given in Table 54 should be multiplied by the improved nail lateral modification factor, K_{44}, which has a value 1.2, i.e. $K_{44} = 1.2$.

For the basic loads in Table 54 to apply, the nails should fully penetrate the tabulated standard values in both the headside and pointside members. In addition, for joints where nails are subjected to multiple shear lateral loads, the thickness of the inner member should be at least 0.85 times the standard thickness given in Table 54. Otherwise, the basic shear load should be multiplied by the smaller of the following ratios, where relevant:

(1) actual to standard thickness of headside member, or
(2) actual penetration to standard pointside thickness, or
(3) actual inner member to 0.85 times the standard thickness.

It is important to note that no load-carrying capacity should be assumed where the ratios given above are less than 0.66 for softwoods and 1.0 for hardwoods. Where improved nails are used, the ratios may be reduced to 0.50 for softwoods and 0.75 for hardwoods. In addition no increase in basic load is permitted for thicknesses or penetrations greater than the standard values.

Therefore, with reference to Fig. 9.6, it is possible to define a modification factor for thickness of members and penetration depth (i.e. bearing lengths) for a nailed connection, $K_{nail-bearing}$, as:

for a nail in single shear,

$$K_{nail-bearing} = \left(\text{the lesser of } \frac{t_{a,head}}{t_{standard}}, \frac{t_{a,point}}{t_{standard}} \text{ or } 1 \right)$$

for a nail in multiple shear,

$$K_{nail-bearing} = \left(\text{the lesser of } \frac{t_{a,head}}{t_{standard}}, \frac{t_{a,inner}}{0.85 \cdot t_{standard}}, \frac{t_{a,point}}{t_{standard}} \text{ or } 1 \right)$$

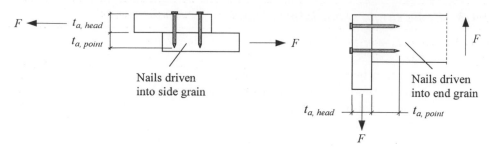

(a) Nails in single shear

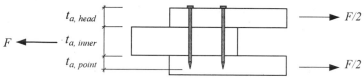

(b) Nails in double shear

Fig. 9.6   Member thicknesses in nailed joints.

2.   Steel plate-to-timber nailed joints

Steel plate-to-timber nailed joints can have a considerably higher load capacity than timber-to-timber joints with similar nailing arrangements. BS 5268 : Part 2, Clause 6.4.5 recommends a 25% higher load capacity for steel plate-to-timber joints provided that steel plates have a minimum thickness equal to the greater of 1.2 mm or 0.3 times the nail diameter.

Where a pre-drilled steel plate or component of such thickness is nailed to a timber member, the basic single shear lateral load given in Table 54 should be multiplied by the steel-to-timber modification factor, K_{46}, which has a value of 1.25, i.e. $K_{46} = 1.25$.

It is recommended that the diameter of the hole in the steel plate should not be greater than the diameter of the nail.

3.   Plywood-to-timber nailed joints

The basic single shear lateral loads for single nails in a plywood-to-timber joint, where the nails are driven through the plywood at right angles into the side grain of timber in service classes 1 and 2, are given in Table 56 of the code (reproduced here as Table 9.6).

For the basic loads in Table 56 of the code to apply, the plywood should be selected from those given in Section 4 of the code, and the nails should be fully embedded and should have an overall length not less than those tabulated.

4.   Tempered hardboard-to-timber nailed joints

The basic single shear lateral loads for single nails in a tempered hardboard-to-timber joint, where the nails are driven through the tempered hardboard

Table 9.6 Basic single shear lateral loads for round wire nails in a plywood-to-timber joint, (Table 56, BS 5268 : Part 2)

Plywood	Nail		Basic single shear lateral load, (N)											
			Strength class											
			Softwoods (not pre-drilled)								Hardwoods (pre-drilled)			
Nominal thickness (mm)	Diameter (mm)	Length (mm)	C14		C16/18/22		C24		C27/30/35/40		D30/35/40		D50/60/70	
			Plywood group[a]								Plywood group[a]			
			I	II	I	II	I	II	I	II	I	II	I	II
6	2.65	40	207	216	212	221	220	230	224	234	267	281	270	296
	3.00	50	252	262	258	268	268	279	273	284	295	346	295	358
	3.35	50	301	311	308	319	318	332	318	339	318	387	318	387
	3.75	75	344	373	344	382	344	399	344	407	344	419	344	419
	4.00	75	360	414	360	424	360	438	360	438	360	438	360	438
9	2.65	40	218	230	223	241	231	250	235	255	274	301	286	315
	3.00	50	261	280	267	286	277	297	281	302	333	362	347	379
	3.35	50	308	328	315	335	327	348	332	355	397	429	415	450
	3.75	75	366	387	374	396	389	412	395	419	479	514	495	539
	4.00	175	405	427	414	436	430	454	438	463	518	571	518	599
12	2.65	45	240	256	244	268	253	278	256	282	297	330	309	345
	3.00	50	281	306	287	312	297	324	302	329	353	389	367	406
	3.35	50	320	336	334	353	345	374	351	380	415	454	432	474
	3.75	75	383	410	391	419	406	435	412	443	492	535	513	559
	4.00	75	421	449	430	459	446	477	453	485	545	590	567	616
15	2.65	50	266	285	271	298	280	310	284	314	327	365	339	380
	3.00	65	308	337	314	343	324	356	329	361	382	424	397	441
	3.35	65	353	383	360	391	372	405	378	412	442	487	459	508
	3.75	75	408	440	416	449	431	466	438	473	518	566	538	590
	4.00	75	445	478	454	488	470	506	477	515	568	619	591	645
18	2.65	50	272	292	283	301	305	314	311	320	361	388	374	409
	3.00	65	339	361	345	369	356	386	361	393	417	461	432	479
	3.35	65	384	417	391	425	404	440	410	446	477	525	495	546
	3.75	75	439	474	447	483	462	500	469	508	552	603	572	627
	4.00	75	476	511	485	522	501	540	509	549	601	655	623	681
21	2.65	50	275	280	288	2C3	306	314	311	320	372	387	391	408
	3.00	65	352	360	360	369	375	385	382	393	456	479	472	504
	3.35	65	411	432	426	443	440	462	446	471	517	570	535	592
	3.75	75	474	513	483	523	499	541	506	549	591	649	612	674
	4.00	75	510	551	520	562	537	581	545	590	641	701	664	727
29	2.65	50	233	235	242	245	261	263	270	273	372	380	391	399
	3.00	65	352	356	360	365	375	380	382	388	461	470	483	494
	3.35	65	403	407	423	426	450	456	458	465	557	568	583	597
	3.75	75	509	515	521	527	542	550	552	560	675	690	707	723
	4.00	75	552	572	573	586	603	611	614	623	754	770	789	808

[a] Plywood group I comprises:
(a) American construction and industrial plywood
(b) Canadian Douglas fir plywood
(c) Canadian softwood plywood
(d) Finnish conifer plywood
(e) Swedish softwood plywood
Plywood group II comprises:
(a) Finnish birch-faced plywood
(b) Finnish birch plywood

at right angles into the side grain of timber in service classes 1 and 2, are given in Table 57 of the code.

For the basic loads in Table 57 of the code to apply, the tempered hardboard should be selected from those given in Section 5 of the code, and the nails should be fully embedded and should have an overall length not less than those tabulated.

5.   Particleboard-to-timber nailed joints
The basic single shear lateral loads for single nails in a particleboard-to-timber joint, where the nails are driven through the particleboard at right angles into the side grain of timber in service classes 1 and 2, are given in Table 58 of the code.

For the basic loads in Table 58 to apply, the particleboard should be selected from those given in Section 9 of the code, and the nails should be fully embedded and should have an overall length not less than:

(1)   2.5 times the nominal particleboard thickness for particleboards in the range 6 mm to 19 mm thick, and
(2)   2.0 times the nominal particleboard thickness for particleboards in the range 20 mm to 40 mm thick.

9.7.4 *Axially loaded nails (withdrawal loads)*

Round wire nails are weak when loaded axially and, thus, their strength should not be relied on to any great extent. The best resistance is generally obtained when the nails are driven into side grain (Fig. 9.7). No withdrawal load is permitted by the code for nails driven into the end grain of timber.

Some of the main factors that influence the withdrawal resistance of a nail include the density and moisture content of the timber into which the nail is

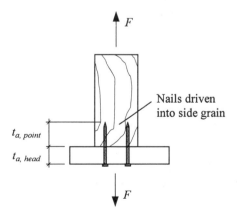

Fig. 9.7   Axially loaded nails.

Table 9.7 Basic withdrawal loads per millimetre of pointside penetration for smooth round wire nails driven at right angles to the grain (Table 55, BS 5268 : Part 2)

Nail diameter (mm)	Basic withdrawal load[a], (N/mm)					
	Strength class					
	C14	C16/18/22	C24	C27/30/35/40	D30/35/40	D50/60/70
2.65	1.28	1.50	1.99	2.42	5.86	9.80
3.00	1.45	1.69	2.25	2.74	6.64	11.09
3.35	1.62	1.89	2.52	3.06	7.41	12.39
3.75	1.81	2.12	2.82	3.42	8.30	13.87
4.00	1.93	2.26	3.00	3.65	8.85	14.79
4.50	2.17	2.54	3.38	4.11	9.96	16.64
5.00	2.41	2.82	3.76	4.56	11.06	18.49
5.60	2.70	3.16	4.21	5.11	12.39	20.71
6.00	2.90	3.39	4.51	5.48	13.28	22.19
6.70	3.23	3.78	5.03	6.12	14.82	24.78
8.00	3.86	4.51	6.01	7.30	17.70	29.58

[a] The pointside penetration of the nail should not be less than 15 mm

driven and the nail type. As mentioned earlier, the type and surface condition of nails enhances their performance, for example, cement or polymer coated, annular ringed shank and helical threaded shank and square twisted nails all perform better under withdrawal loading than round wire nails. In addition such improved nails are less affected by changes in the moisture content of the timber.

The basic withdrawal loads for single nails driven at right angles to the side of timber in service classes 1 and 2 are given in Table 55 of the code (reproduced here as Table 9.7). The values apply to each 1 mm depth of actual pointside penetration achieved; where the pointside penetration of the nail should not be less than 15 mm.

For the threaded part of annular ringed shank and helical threaded shank nails, BS 5268 : Part 2 recommends that the basic withdrawal loads given in Table 55 should be multiplied by the improved nail withdrawal modification factor, K_{45}, which has a value of 1.5, i.e. $K_{45} = 1.5$.

9.7.5 *Permissible load for a nailed joint*

Clause 6.4.9 of BS 5268 : Part 2 states that the permissible load for a nailed joint should be determined as the sum of the permissible loads for each nail in the joint, where each permissible nail load, F_{adm}, should be calculated from the equation:

$$F_{adm} = F \times K_{48} K_{49} K_{50} \tag{9.3}$$

where:

F is the basic load for a nail under shear loading (Table 54) or withdrawal loading (Table 55) and is taken from Clauses 6.4.4, 6.4.5.2,
6.4.6.2, 6.4.7.2 or 6.4.8.2 of the code which were described earlier.
Therefore F is given by:

(a)   for a nail under single or multiple shear lateral load,

$$F = F_{basic} \text{ (Table 54)} \times n_{shear} K_{pre-drill} K_{nail-bearing} K_{43} K_{44} K_{46}$$
$$(9.4)$$

(b)   for a nail under axial (withdrawal) load with a pointside penetration depth of t_{point},

$$F = F_{basic} \text{ (Table 55)} \times t_{point} K_{45} \qquad (9.5)$$

where:

$n_{shear} =$ number of shear planes per nail

$K_{pre-drill} = 1.15$, the pre-drilling factor for softwoods in strength
classes C14 to C40 and TR26

$K_{nail-bearing} = \left(\text{the lesser of } \dfrac{t_{a,head}}{t_{standard}}, \dfrac{t_{a,point}}{t_{standard}} \text{ or } 1 \right)$ for a nail in
single shear

$= \left(\text{the lesser of } \dfrac{t_{a,head}}{t_{standard}}, \dfrac{t_{a,inner}}{0.85 \cdot t_{standard}}, \dfrac{t_{a,point}}{t_{standard}} \text{ or } 1 \right)$

for a nail in multiple shear

$K_{43} = 0.7$, for nailing into the end grain

$K_{44} = 1.2$, improved nail shear lateral modification factor

$K_{45} = 1.5$, improved nail withdrawal modification factor

$K_{46} = 1.25$, steel plate-to-timber modification factor.

K_{48} is the modification factor for duration of loading

where:

$K_{48} = 1.00$, for long-term loads

$= 1.25$, for tempered hardboard-to-timber joints for medium-
term loads

$= 1.4$, for particleboard-to-timber joints for medium-term loads

$= 1.12$, for other than tempered hardboard-to-timber and particleboard-to-timber joints for medium-term loads

$= 1.62$, for tempered hardboard-to-timber joints for short and
very short-term loads

$= 2.10$, for particleboard-to-timber joints for short and very short-
term loads

$= 1.25$, for other than tempered hardboard-to-timber and particleboard-to-timber joints for short and very short-term loads.

K_{49} is the modification factor for moisture content

where:
$K_{49} = 1.00$, for lateral loads in joints in service classes 1 and 2
$\quad\;\; = 0.70$, for lateral loads in timber-to-timber joints in service class 3
$\quad\;\; = 1.00$, for lateral loads using annular ringed shank nails and helical threaded shank nails in all service class conditions
$\quad\;\; = 1.00$, for withdrawal load in all constant service class conditions
$\quad\;\; = 0.25$, for withdrawal loads where cyclic changes in moisture content can occur after nailing.

K_{50} is the modification factor for the number of nails of the same diameter in each line, acting in shear, for an axially loaded member, loaded parallel to the line of nails

where:
$K_{50} = 1.00$, for the number of nails in each line < 10
$\quad\;\; = 0.9$, for the number of nails in each line ≥ 10
$\quad\;\; = 1.00$, for all other loading cases.

9.8   Screwed joints

Wood screws are used in place of nails in applications requiring higher capacities, in particular in situations where a greater resistance to withdrawal is required. In general withdrawal resistance of screws is 2 to 3 times that of nails of a similar size. They can be used for timber-to-timber joints but they are especially suitable for steel-to-timber and panel-to-timber joints.

Screws should always be inserted by threading into pre-drilled holes in the timber using a screwdriver and not hammered, and the basic values given in the code are based on this assumption. Otherwise the load-carrying capacity, mainly the withdrawal resistance, will decrease significantly. The hole for the shank should have a diameter equal to the shank diameter and should have the same length as the shank. The pilot hole for the threaded portion of the screw should have a diameter of about half the shank diameter.

The most common types of wood screws are the countersunk head, round head and coach screws. These are illustrated in Fig. 9.8.

The design recommendations given in BS 5268 : Part 2, Clause 6.5 apply to steel screws which conform to BS 1210 : 1963. The diameter relates to the smooth shank, which ranges from 3.45–7.72 mm for countersunk and round headed screws. Coach screws, in general, range from 6–20 mm in shank diameter.

Coach screws, with nut-like head, require a washer and should always be threaded into pre-drilled holes. They are available in lengths of 25–300 mm and are most suitable in large connections to hold timber connectors such as dowels in place or in situations where the use of bolts is not suitable due to one-sided access.

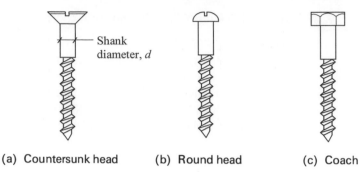

(a) Countersunk head   (b) Round head   (c) Coach

Fig. 9.8 Common types of wood screws.

9.8.1 *Basic single shear lateral loads*

1. Timber-to-timber screwed joints

The basic single shear lateral loads for single screws with a minimum tensile strength of $550\,N/mm^2$, threaded at right angles to the side grain of timber in service classes 1 and 2, are given in Table 60 of the code (reproduced here as Table 9.8).

For screws inserted into the end grain of timber, the values given in Table 60 should be multiplied by the end grain modification factor, K_{43}, which has a value of 0.7, i.e. $K_{43} = 0.7$.

For the basic loads in Table 60 to apply, the headside member thickness should be equal to the value given in that table and the penetration of the screw in the pointside member should be at least twice the actual headside member thickness. Otherwise, the basic shear load should be multiplied by the ratio of the actual headside thickness to the standard thickness given in Table 60, provided that the pointside screw penetration is at least twice the actual headside thickness.

Table 9.8 Basic single shear lateral loads for screws inserted into pre-drilled holes in a timber-to-timber joint (Table 60, BS 5268 : Part 2)

Screw		Standard headside thickness (mm)	Basic single shear lateral load, (N)					
			Strength class					
Number	Shank diameter (mm)		C14	C16/18/22	C24	C27/40/35/40	D30/35/40	D50/60/70
6	3.45	12	253	271	303	314	396	455
8	4.17	16	390	407	440	456	584	677
10	4.88	22	546	572	622	647	846	993
12	5.59	35	819	863	952	985	1179	1306
14	6.30	38	1002	1056	1163	1217	1457	1613
16	7.01	44	1251	1319	1429	1470	1759	1948
18	7.72	47	1469	1549	1694	1742	2085	2308

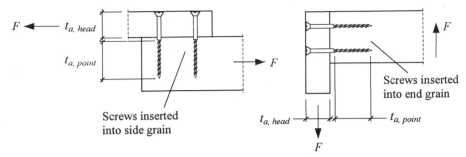

Fig. 9.9   Member thicknesses in screwed joints.

Therefore, with reference to Fig. 9.9, it is possible to define a modification factor for thickness of the headside member (i.e. bearing length) of a screwed connection, $K_{screw\text{-}bearing}$, as:

$$K_{screw\text{-}bearing} = \frac{t_{a,head}}{t_{standard}}$$

The minimum headside member thickness should not be less than twice the shank diameter. In addition no increase in basic load is permitted for thicknesses or penetrations greater than the standard values.

2. Steel plate-to-timber screwed joints
BS 5268 : Part 2, Clause 6.5.5 recommends a 25% higher load capacity for steel plate-to-timber screwed joints, provided that steel plates have a minimum thickness equal to the greater of 1.2 mm or 0.3 times the shank diameter of the screw.

Where a pre-drilled steel plate or component of such thickness is screwed to a timber member, the basic single shear lateral load given in Table 60 should be multiplied by the steel-to-timber modification factor, K_{46}, which has a value of 1.25, i.e. $K_{46} = 1.25$.

3. Plywood-to-timber screwed joints
The basic single shear lateral loads for single screws in a plywood-to-timber joint, where the screws of minimum length are threaded through the plywood at right angles into the side grain of timber in service classes 1 and 2, are given in Table 62 of the code.

For the basic loads in Table 62 to apply, the plywood should be selected from those given in the Section 4 of the code and the diameter of the pre-drilled hole in the plywood should not be greater than the shank diameter of the screw.

9.8.2   *Axially loaded screws (withdrawal loads)*

The basic withdrawal loads for single screws inserted at right angles to the side of timber in service classes 1 and 2 are given in Table 61 of the code

Table 9.9 Basic withdrawal loads per millimetre of pointside penetration for screws turned at right angles to grain (Table 61, BS 5268 : Part 2)

Screw		Basic withdrawal load[a], (N/mm)					
				Strength class			
Number	Shank diameter (mm)	C14	C16/18/22	C24	C27/40/35/40	D30/35/40	D50/60/70
6	3.45	10.96	12.03	14.29	16.06	27.31	37.17
8	4.17	12.63	13.87	16.47	18.51	31.49	42.85
10	4.88	14.21	15.61	18.53	20.83	35.43	48.22
12	5.59	15.73	17.28	20.52	23.06	39.23	53.39
14	6.30	17.21	18.90	22.44	25.22	42.91	58.40
16	7.01	18.64	20.48	24.31	27.33	46.49	63.26
18	7.72	20.04	22.01	26.14	29.38	49.97	68.01

[a] The penetration of the screw point should not be less than 15 mm

(reproduced here as Table 9.9). The values apply to each 1 mm depth of actual pointside penetration achieved by the threaded part of the screw; where the penetration of the screw point should not be less than 15 mm.

No withdrawal load is permitted by the code for screws inserted into the end grain of timber.

9.8.3 Permissible load for a screwed joint

Clause 6.5.7 of BS 5268 : Part 2 recommends that the permissible load for a screwed joint should be determined as the sum of the permissible loads for each screw in the joint, where each permissible screw load, F_{adm}, should be calculated from the equation:

$$F_{adm} = F \times K_{52}K_{53}K_{54} \tag{9.6}$$

where:

F is the basic load for a screw under shear (Tables 60 or 62) or withdrawal loading (Table 61) and is taken from Clauses 6.5.4, 6.5.5.2 and 6.5.6.2 of the code, which were described earlier. Therefore F is given by:

for a screw under shear lateral load,

$$F = F_{basic} \text{ (Tables 60 or 62)} \times K_{screw\text{-}bearing}K_{43}K_{46} \tag{9.7}$$

for a screw under axial (withdrawal) load with a pointside threaded penetration depth of t_{point},

$$F = F_{basic} \text{ (Table 61)} \times t_{point} \tag{9.8}$$

where:

$$K_{screw\text{-}bearing} = \frac{t_{a,head}}{t_{standard}}$$

$K_{43} = 0.7$, for screwing into the end grain

$K_{46} = 1.25$, steel plate-to-timber modification factor.

K_{52} is the modification factor for duration of loading

where:

$K_{52} = 1.00$, for long-term loads

$\phantom{K_{52}} = 1.12$, for medium-term loads

$\phantom{K_{52}} = 1.25$, for short and very short-term loads.

K_{53} is the modification factor for moisture content

where:

$K_{53} = 1.00$, for a joint in service classes 1 and 2

$\phantom{K_{53}} = 0.70$, for a joint in service class 3.

K_{54} is the modification factor for the number of screws of the same diameter in each line, acting in shear, for an axially loaded member, loaded parallel to the line of screws

where:

$K_{54} = 1.00$, for the number of screws in each line < 10

$\phantom{K_{54}} = 0.9$, for the number of screws in each line ≥ 10

$\phantom{K_{54}} = 1.00$, for all other loading cases.

9.9   Bolted and dowelled joints

Dowels are cylindrical rods generally made of steel with a smooth surface, and are available in diameters ranging from 6–30 mm. Bolts are threaded dowels with hexagonal or semi-sphere heads and hexagonal nuts. Dowels and bolts are commonly used in connections requiring higher shear capacity than the nailed or screwed joints, see Fig. 9.10. Threaded dowels with nuts and bolts are also used in axially loaded tension connections.

Both dowels and bolts are commonly used to hold two or more members together, but the most common connection type involves three or more members in a multiple shear arrangement. The side members can be either timber or steel. When bolts are used, washers are required under the bolt head and under the nut to distribute the loads. The recommended washer diameter is 3 times the bolt diameter ($3d$) with a thickness of 0.25 times the bolt diameter ($0.25d$). When tightened, a minimum of one complete thread should protrude from the nut.

Dowelled joints are easy to fabricate. They are inserted into pre-drilled holes with a diameter not greater than the dowel diameter itself. They can produce stiffer joints compared to bolted ones which require 1 mm oversize

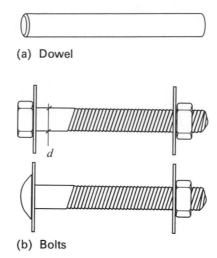

(a) Dowel

(b) Bolts

Fig. 9.10   Typical dowel and bolts.

for clearance; BS 5268 : Part 2 recommends a maximum of 2 mm oversizing for bolt holes. If steel plates are incorporated in a dowelled connection, a 1 mm clearance must be provided and due allowance should then be made for the additional deformation that may occur.

9.9.1   *Basic single shear lateral loads*

1.   Timber-to-timber bolted or dowelled joints
The basic single shear lateral loads for a single bolt or dowel, with a minimum tensile strength of 400 N/mm^2, in a *two-member* timber joint, in which the load acts perpendicular to the axis of the bolt or dowel, and parallel or perpendicular to the grain of the timber, are given in Tables 63–68 of BS 5268 : Part 2. Table 64 is reproduced here as Table 9.10.

The basic loads appropriate to each shear plane in a *three-member* joint under the same conditions are given in Tables 69–74 of the code. Table 70 is reproduced here as Table 9.11. Separate loads are tabulated for long-, medium- and short-duration loads and therefore there is no need to further modify the basic load values for duration of loading.

Unlike nails and screws, bolts and dowels have loading capacities that vary depending on the direction of the applied load with respect to the grain direction of the timber. This is why the code provides separate basic shear values for loadings parallel and perpendicular to the grain of timber, and recommends the use of Hankinson's equation for determination of basic loads, F, in situations where the applied load is inclined at an angle α to the grain of the timber, see Fig. 9.11, hence:

$$F = \frac{F_{/\!/} F_{\perp}}{F_{/\!/} \sin^2 \alpha + F_{\perp} \cos^2 \alpha} \tag{9.9}$$

Table 9.10 Basic single shear loads for one 4.6 grade steel bolt or dowel in a two-member timber-to-timber joint: C16/18/22 timber (Table 64, BS 5268 : Part 2)

Load duration	Minimum member thickness (mm)	C16/18/22									
		Basic single shear load, (kN)									
		Direction of loading									
		Parallel to the grain Bolt or dowel diameter, (mm)					Perpendicular to the grain Bolt or dowel diameter, (mm)				
		M8	M12	M16	M20	M24	M8	M12	M16	M20	M24
Long-term	16	0.32	0.46	0.58	0.70	0.79	0.29	0.40	0.49	0.56	0.61
	22	0.44	0.63	0.80	0.96	1.09	0.40	0.55	0.67	0.77	0.84
	35	0.70	1.00	1.28	1.52	1.73	0.63	0.87	1.06	1.22	1.34
	44	0.88	1.26	1.61	1.91	2.18	0.79	1.09	1.34	1.54	1.69
	47	0.94	1.35	1.72	2.04	2.33	0.85	1.17	1.43	1.64	1.80
	60	1.20	1.72	2.19	2.61	2.97	1.08	1.49	1.83	2.09	2.30
	72	1.44	2.07	2.63	3.13	3.57	1.30	1.79	2.19	2.51	2.76
	97	1.47	2.78	3.54	4.22	4.81	1.39	2.41	2.95	3.39	3.72
	147	1.47	3.23	5.37	6.39	7.29	1.39	3.00	4.47	5.13	5.64
Medium-term	16	0.41	0.59	0.75	0.89	1.02	0.37	0.51	0.63	0.72	0.79
	22	0.57	0.81	1.03	1.23	1.40	0.51	0.70	0.86	0.99	1.09
	35	0.90	1.29	1.64	1.96	2.23	0.81	1.12	1.37	1.57	1.73
	44	1.13	1.62	2.07	2.46	2.80	1.02	1.41	1.72	1.97	2.17
	47	1.21	1.73	2.21	2.63	2.99	1.09	1.50	1.84	2.11	2.32
	60	1.54	2.21	2.82	3.35	3.82	1.39	1.92	2.35	2.69	2.96
	72	1.63	2.66	3.38	4.02	4.59	1.55	2.30	2.82	3.23	3.55
	97	1.63	3.58	4.55	5.42	6.18	1.55	3.10	3.79	4.35	4.79
	147	1.63	3.59	6.24	8.22	9.37	1.55	3.34	5.69	6.60	7.26
Short-term and very short-term	16	0.46	0.66	0.55	1.01	1.15	0.42	0.57	0.70	0.81	0.89
	22	0.64	0.91	1.16	1.38	1.58	0.57	0.79	0.97	1.11	1.22
	35	1.01	1.45	1.85	2.20	2.51	0.91	1.26	1.54	1.77	1.94
	44	1.27	1.83	2.32	2.77	3.15	1.15	1.58	1.94	2.22	2.44
	47	1.36	1.95	2.48	2.96	3.37	1.23	1.69	2.07	2.37	2.61
	60	1.73	2.49	3.17	3.77	4.30	1.56	2.16	2.64	3.03	3.33
	72	1.73	2.99	3.80	4.53	5.16	1.64	2.59	3.17	3.64	4.00
	97	1.73	3.81	5.12	6.10	6.95	1.64	3.49	4.27	4.90	5.39
	147	1.73	3.81	6.62	9.24	10.54	1.64	3.54	6.04	7.42	8.16

where:

$F_{/\!/}$ = basic load parallel to the grain, obtained from Tables 63–74

$F_{\perp}$ = basic load perpendicular to the grain, obtained from Tables 63–74.

If a load F acts at an angle β to the axis of the bolt the component of the load perpendicular to the axis of the bolt, ($F\sin\beta$), should not be greater

Table 9.11 Basic single shear loads for one 4.6 grade steel bolt or dowel in a three-member timber-to-timber joint: C16/18/22 timber (Table 70, BS 5268 : Part 2)

Load duration	Minimum outer member thickness[a] (mm)	C16/18/22									
		Basic single shear load, (kN)									
		Direction of loading									
		Parallel to the grain Bolt or dowel diameter, (mm)					Perpendicular to the grain Bolt or dowel diameter, (mm)				
		M8	M12	M16	M20	M24	M8	M12	M16	M20	M24
Long-term	16	0.77	1.11	1.41	1.68	1.91	0.70	0.96	1.18	1.35	1.48
	22	1.04	1.52	1.94	2.31	2.63	0.96	1.32	1.62	1.85	2.04
	35	1.10	2.29	3.09	3.67	4.19	1.03	2.10	2.57	2.95	3.24
	44	1.19	2.34	3.88	4.62	5.26	1.10	2.16	3.23	3.71	4.08
	47	1.22	2.36	3.97	4.93	5.62	1.13	2.17	3.45	3.96	4.36
	60	1.38	2.50	4.06	6.06	7.18	1.27	2.28	3.66	5.06	5.56
	72	1.47	2.68	4.20	6.14	8.49	1.39	2.42	3.76	5.45	6.67
	97	1.47	3.12	4.63	6.48	8.72	1.39	2.79	4.08	5.66	7.55
	147	1.47	3.23	5.60	7.63	9.74	1.39	13.00	4.99	16.48	8.20
Medium-term	16	0.99	1.43	1.81	2.16	2.46	0.90	1.23	1.51	1.73	1.91
	22	1.17	1.96	2.49	2.97	3.38	1.10	1.70	2.08	2.38	2.62
	35	1.28	2.58	3.97	4.72	5.38	1.20	2.38	3.31	3.79	4.17
	44	1.40	2.67	4.44	5.94	6.77	1.30	2.45	4.04	4.77	5.24
	47	1.45	2.71	4.47	6.34	7.23	1.34	2.48	4.05	5.09	5.60
	60	1.63	2.94	4.63	6.81	9.23	1.54	2.66	4.16	6.05	7.15
	72	1.63	3.19	4.87	6.98	9.53	1.55	2.87	4.32	6.15	8.33
	97	1.63	3.59	5.51	7.55	9.97	1.55	3.34	4.81	6.53	8.56
	147	1.63	3.59	6.24	9.22	11.55	1.55	3.34	5.69	7.76	9.61
Short-term and very short-term	16	1.12	1.60	2.04	2.43	2.77	1.01	1.39	1.70	1.95	2.15
	22	1.24	2.20	2.81	3.34	3.81	1.17	1.91	2.34	2.68	2.95
	35	1.39	2.75	4.46	5.31	6.06	1.29	2.54	3.72	4.27	4.69
	44	1.53	2.87	4.73	6.68	7.62	1.42	2.63	4.29	5.36	5.90
	47	1.59	2.92	4.77	7.14	8.13	1.47	2.67	4.31	5.73	6.30
	60	1.73	3.19	4.98	7.27	10.04	1.64	2.88	4.46	6.44	8.05
	72	1.73	3.49	5.27	7.49	10.17	1.64	3.13	4.66	6.58	8.86
	97	1.73	3.81	6.02	8.18	10.73	1.64	3.54	5.25	7.05	9.18
	147	1.73	3.81	6.62	10.09	12.62	1.64	3.54	6.04	8.50	10.45

[a] The corresponding minimum inner member thickness is assumed to be double the tabulated value

than the basic load given in Tables 63–74, modified where appropriate by Hankinson's equation, see Fig. 9.12.

Clause 6.6.4.1 of the code specifies that for two-member joints, where parallel members are of unequal thickness, the load for the thinner member

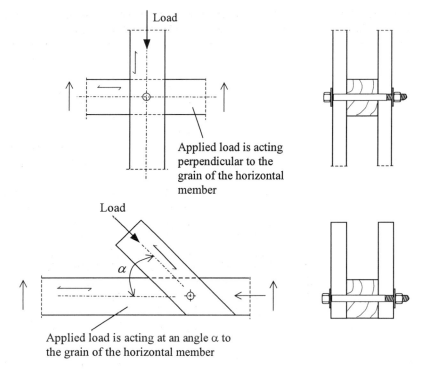

Fig. 9.11   Load at an angle to the grain.

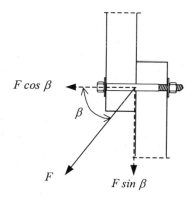

Fig. 9.12   Load at an angle to the axis of the bolt.

should be used. Where members of unequal thickness are joined at an angle, the basic load for each member should be determined, using Hankinson's equation, and the smaller load should be used.

For three-member joints, the basic loads given in Tables 69–74 apply to joints where the outer members have the tabulated thickness and the inner member is twice as thick. For other thicknesses the load should be taken as the lesser of the loads obtained by linear interpolation between the two adjacent tabulated thicknesses for each member.

The basic load for a joint with more than three members should be taken as the sum of the basic loads for each shear plane, assuming that the joint consists of a series of three-member joints.

2. Steel plate-to-timber bolted or dowelled joints
BS 5268 : Part 2, Clause 6.6.5.1 recommends a 25% higher load capacity for steel plate-to-timber bolted or dowelled joints, provided that steel plates have a minimum thickness equal to the greater of 2.5 mm or 0.3 times the bolt or dowel diameter.

Where a steel plate or component of such thickness is bolted or dowelled to a timber member, the basic loads given in Tables 63–74 should be multiplied by the steel-to-timber modification factor, K_{46}, which has a value of 1.25, i.e. $K_{46} = 1.25$.

In dowelled joints where the steel plates are used as the side (outer) members of the joint, the steel plates should be secured in position adequately, i.e. by threading and applying nuts to the dowel ends or by welding dowel ends to the steel plates.

No increase should be made to the basic load when the timber is loaded perpendicular to the grain.

9.9.2 *Permissible load for a bolted or dowelled joint*

Clauses 6.6.6 and 6.6.7.4 of BS 5268 : Part 2 state that the permissible load for bolted or dowelled joints should be determined as the sum of the permissible loads for each bolt or dowel in the joint, where each permissible bolt or dowel load, F_{adm}, should be calculated from the equation:

$$F_{adm} = F \times K_{56}K_{57} \tag{9.10}$$

where:
F is the basic load for a bolt or dowel under lateral loading per shear plane (Tables 63–74) and is taken from Clauses 6.6.4, 6.5.5 and 6.6.7 of the code which were described earlier. Therefore F is given by:

(a)   for a bolt or dowel in a two-member joint,

$$F = F_{basic} \text{ (Tables 63–68)} \times K_{46} \tag{9.11}$$

(b)   for a bolt or dowel in a symmetrical three-member joint

$$F = F_{basic} \text{ (Tables 69–74)} \times 2 \times K_{46} \tag{9.12a}$$

(c)   in general, for a bolt or dowel in a joint with three or more members,

$$F = \sum F_{basic} \text{ (Tables 69–74)} \times K_{46} \tag{9.12b}$$

where:

$\sum F_{basic}$ = sum of the modified basic loads

K_{46} = 1.25, steel plate-to-timber modification factor.

It is to be noted that in all cases, if load is applied at an angle α to the grain of timber, F_{basic} is obtained from

$$F_{basic} = \frac{F_{//} F_{\perp}}{F_{//} \sin^2 \alpha + F_{\perp} \cos^2 \alpha}$$

K_{56} is the modification factor for moisture content

where:

K_{56} = 1.00, for a joint in service classes 1 and 2

= 0.70, for a joint made in timber of service class 3 and used in that service class

= 0.40, for a joint made in timber of service class 3 but used in service classes 1 and 2.

K_{57} is the modification factor for the number of bolts or dowels of the same diameter in each line acting in shear for an axially loaded member, loaded parallel to the line of bolts or dowels, where:

$$K_{57} = 1 - \frac{3(n-1)}{100} \qquad \text{for } n < 10$$

$$= 0.7 \qquad \text{for } n \geq 10$$

$$= 1.0 \qquad \text{for all other loading cases}$$

where n is the number of bolts or dowels in each line.

9.10   Moment capacity of dowel-type fastener joints

Structural connections are often required to transmit moments as well as shear and/or axial forces. Figure 9.13 presents some examples of moment-resisting timber joints such as (a) a splice joint in a continuous beam, (b) a moment-resisting column base and (c) a knee joint in a frame. Although in this section a bolted connection is used to illustrate a simplified analysis method for determination of connection moment capacity, the method can similarly be applied to nailed, screwed, or, indeed, dowelled joints.

Figure 9.14(a) shows a bolted connection which is subjected to a moment of M, and shear and normal forces of V and H. The effects of the moment M and the forces V and H, all acting at the geometric centre of the bolt group at C, can be considered separately.

First consider the effect of the applied moment, M. For an elastic behaviour of the connection, we can assume that the moment, M, will induce a shearing force, F_M, in any of the bolts perpendicular to the line

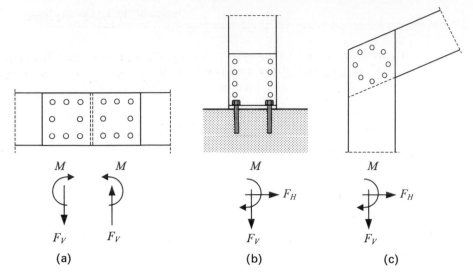

Fig. 9.13 Moment-resisting connections.

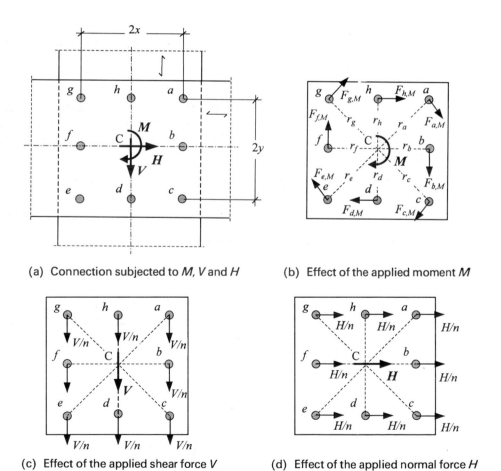

(a) Connection subjected to *M*, *V* and *H*

(b) Effect of the applied moment *M*

(c) Effect of the applied shear force *V*

(d) Effect of the applied normal force *H*

Fig. 9.14 Moment-resisting joint subjected to an applied moment and shear and normal forces.

joining C to the bolt. Therefore the force F_M will be perpendicular to the distance r from the bolt to C (see Fig. 9.14(b)). For equilibrium we have:

$$M = \sum F_M \cdot r \qquad (9.13)$$

If we consider $F_M = k \cdot r$, where k is constant for all bolts, then

$$M = k \sum r^2 \qquad (9.14)$$

Thus,

$$k = \frac{M}{\sum r^2}$$

By substituting for k, the force acting on a bolt due to the moment, M, becomes:

$$F_M = \frac{M}{\sum r^2} r \qquad (9.15)$$

where: $\sum r^2 = r_a^2 + r_b^2 + r_c^2 + r_d^2 + r_e^2 + \cdots$.

As an example, the force acting on the bolt a due to the applied moment M is:

$$F_{a,M} = \frac{M}{\sum r^2} r_a$$

Now considering the effects of shear and normal forces V and H, assume that each one is distributed equally among the bolts as a shearing force parallel to the line of action of V and/or H (see Fig. 9.14(c), (d)). For a joint with n number of bolts,

the force exerted on a bolt due to the applied shear force V is,

$$F_V = \frac{V}{n} \qquad (9.16)$$

the force exerted on a bolt due to the applied normal force H is,

$$F_H = \frac{H}{n} \qquad (9.17)$$

The total force acting on a bolt can then be calculated as the vectorial summation of F_M, F_V and F_H. For example, the total force acting on the bolt a, F_a, with its angle of inclination to the grain of timber, α, is obtained as illustrated in Fig. 9.15.

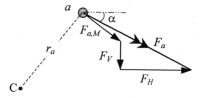

Fig. 9.15   Vectorial summation of $F_{a,M}$, F_V and F_H acting on the bolt a.

With forces defined by equations 9.15–9.17, moment resisting joints, with rectangular pattern, can be designed so that the load on the furthest fastener (i.e. fastener a) is not greater than the permissible load value. The total force, F_a, acting on the furthest fastener a can be calculated from:[1]

$$F_a = \sqrt{\left\{\left[F_V + \frac{x}{\sqrt{(x^2 + y^2)}} F_{a,M}\right]^2 + \left[F_H + \frac{y}{\sqrt{(x^2 + y^2)}} F_{a,M}\right]^2\right\}} \quad (9.18)$$

acting at an angle α to the grain of timber where:

$$\alpha = \arctan\left[\frac{xF_{a,M} + \sqrt{(x^2 + y^2)}F_V}{yF_{a,M} + \sqrt{(x^2 + y^2)}F_H}\right] \quad (9.19)$$

A moment-resisting connection with a circular pattern, such as that of joint (c) shown in Fig. 9.13, can be designed considering the load on the fastener located on the longitudinal axis. In such cases the total force on the fastener is given by:

$$F = \sqrt{[(F_V + F_M)^2 + F_H^2]} \quad (9.20)$$

acting at an angle α where:

$$\alpha = \arctan\left(\frac{F_M + F_V}{F_H}\right) \quad (9.21)$$

It is to be noted that for connections with parallel members, Hankinson's formula (equation 9.9) should be used for the calculation of the permissible loads for bolts or dowels loaded at an angle α to the grain of timber. For timber-to-timber connections with members perpendicular to each other, a modified version of Hankinson's formula may be used. Therefore:

for a timber-to-timber bolted or dowelled connection with parallel members or steel-to-timber, the basic single shear load is given by:

$$F = \frac{F_\parallel F_\perp}{F_\parallel \sin^2 \alpha + F_\perp \cos^2 \alpha} \quad (9.9)$$

for bolted or dowelled connections with perpendicular members,[1]

$$F = \frac{2 \cdot F_\parallel F_\perp}{F_\parallel + F_\perp} \quad (9.22)$$

9.11   Connectored joints

Timber connectors are the most efficient of all mechanical fasteners, as they improve the transfer of loads by increasing the bearing area between the fasteners and the timber. They are generally in the form of rings, discs or plates, partially embedded in the faces of adjacent timber members, to transmit the load from one to the other. There are several types of timber connectors of which the most commonly used are toothed-plates, split-rings and shear-plate connectors which require bolts to draw the members together and metal-plate fasteners such as punched metal-plates with integral teeth and nail-plates which allow members to remain in the same plane.

BS 5268 : Part 2 includes rules for the design of joints using bolted timber connectors such as toothed-plates, split-rings and shear-plates, together with recommendations on joint preparation, spacing of connectors and appropriate edge and end distances.

9.11.1 *Toothed-plate connectors*

Toothed-plate connectors are made from hot dipped galvanised or cold rolled mild steel. They are commonly available in circular or square shapes with sizes ranging from 38–165 mm. Toothed-plate connectors are manufactured in either double- or single-sided forms, see Fig. 9.16. Double-sided toothed-plates are suitable for use in timber-to-timber connections, whereas single-sided ones can be used in steel-to-timber connections or in pairs put back-to-back for timber-to-timber joints in situations where it may be necessary to de-mount the connection.[3]

The assembly process involves drilling the bolt hole in the members, then the connector is placed between the timber member(s) and the connection is pressed together, by either a hydraulic press or with the aid of a high-strength bolt and large washers. After pressing, the high-strength bolt is replaced by the permanent black bolt with associated washers. Toothed-plate connectors are not suitable for timber with a density greater than 500 kg/m³.

BS 5268 : Part 2, Clause 6.7 deals with toothed-plate connector joints. Bolt and washer sizes suitable for different shapes and forms of toothed-plate

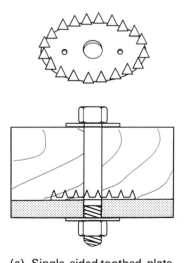

(a) Single-sided toothed-plate   (b) Double-sided toothed-plate

Fig. 9.16 Toothed-plate connectors.

connectors are detailed in Table 76 of the code, and recommendations for the spacing, end and edge distances are given in Tables 77–79.

The basic loads parallel and perpendicular to the grain of the timber for toothed-plate connector units are given in Table 80 of the code. The basic loads apply to softwoods and hardwood, in strength classes C14 to C40, TR26 and D30 to D40, provided that the teeth of the connectors can be fully embedded.

Permissible load for a toothed-plate connected joint

Clause 6.7.6 of BS 5268 : Part 2 recommends that the permissible load for a toothed-plate connectored joint be determined as the sum of the permissible loads for each unit in the joint, where each permissible load, F_{adm}, should be calculated from the equation:

$$F_{adm} = F \times K_{58} K_{59} K_{60} K_{61} \tag{9.23}$$

where:

F is the basic load for a toothed-plate under single shear loading (Table 80)

It is to be noted that if load is applied at an angle α to the grain of timber, F is obtained from

$$F = \frac{F_{/\!/} F_{\perp}}{F_{/\!/} \sin^2 \alpha + F_{\perp} \cos^2 \alpha}$$

K_{58} is the modification factor for duration of loading

     where:
     $K_{58} = 1.00$, for long-term loads
           $= 1.12$, for medium-term loads
           $= 1.25$, for short and very short-term loads.

K_{59} is the modification factor for moisture content

     where:
     $K_{59} = 1.00$, for a joint in service classes 1 and 2
           $= 0.70$, for a joint in service class 3.

K_{60} is the modification factor for end distance

     where:
     $K_{60} = 1.00$ if the standard edge and end distances and spacing given in Tables 77–79 are used, otherwise the lesser of K_c and K_s as given in Tables 81, 82 and 79 of the code should be used.

K_{61} is the modification factor for the number of connector units of the same size in each line for an axially loaded member, loaded parallel to the line of connectors

where:

$$K_{61} = 1 - \frac{3(n-1)}{100} \qquad \text{for } n < 10$$
$$= 0.7 \qquad\qquad \text{for } n \geq 10$$
$$= 1.0 \qquad\qquad \text{for all other loading cases}$$

where n is the number of connector units in each line.

9.11.2   *Split-ring and shear-plate connectors*

Split-ring and shear-plate connectors are manufactured from hot rolled carbon steel, pressed steel, aluminium cast alloy or malleable iron. They require precision grooving and boring in timber members for assembly, and can provide a large shear capacity, much greater than that achievable by toothed-plate connectors.

Split-ring connectors have a split in their ring to provide bearing on both the inner core and the centre surface of the groove and are suitable for timber-to-timber connections only, see Fig. 9.17. They are provided in two sizes of 64 mm and 102 mm diameter and require, respectively, M12 and M20 bolts with associated washers.

BS 5268 · Part 2, Clauses 6.8 deals with split ring connector joints. Bolt and washer sizes and dimensions for suitable circular grooves for split-ring connectors are detailed in Tables 83 and 84 of the code. The recommendations for the end and edge distances and spacings for split-ring connectors for standard dimensions are given in Tables 85–87, and for those less than the standard dimensions in Tables 89 and 90.

The basic loads parallel and perpendicular to the grain of the timber for split-ring connector units are given in Table 88 of the code.

The shear-plate connectors can be used for timber-to-timber joints, in pairs put back-to-back, and also as a single unit for steel-to-timber connections because they can remain flush with the surface of the timber member, see Fig. 9.18. In shear-plate connections, after transfer of the load from one

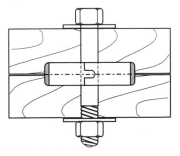

Fig. 9.17   Split-ring connector.

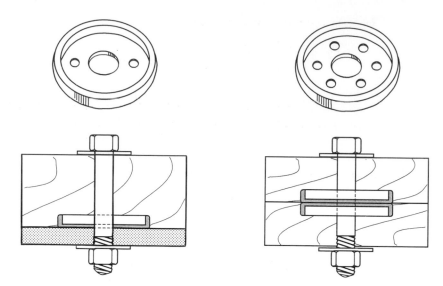

Fig. 9.18 Shear-plate connectors.

member into its connector, the bolt is loaded via bearing stresses between the shear-plate and the bolt, and the load is then transferred into the other member through the second shear-plate or directly into a steel member.

It is to be noted that in all connectored joints, due to a small tolerance required for insertion of the bolt, some initial slip is expected and therefore an allowance should be made for this in design calculations.

BS 5268 : Part 2, Clause 6.9 deals with shear-plate connector joints. Bolt and washer sizes and dimensions for suitable circular grooves for split-ring connectors are detailed in Table 91 and Figure 8 of the code. The recommendations for the end and edge distances and spacings for shear-plate connectors for standard dimensions are given in Tables 85–87, and for less than the standard dimensions in Tables 89 and 90.

The basic loads parallel and perpendicular to the grain of the timber for shear-plate connector units for both timber-to-timber and steel-to-timber joints are given in Table 92 of the code.

Permissible load for split-ring and shear-plate connectored joints

Clauses 6.8.5 and 6.9.6 of BS 5268 : Part 2 state that the permissible load for such connectored joints should be determined as the sum of the permissible loads for each unit in the joint, where each permissible load, F_{adm}, should be calculated from the equation:

for a split-ring connector unit,

$$F_{adm} = F \times K_{62}K_{63}K_{64}K_{65} \qquad (9.24)$$

for a shear-plate connector unit,

$$F_{adm} = F \times K_{66}K_{67}K_{68}K_{69}$$

$$\leq \text{(the limiting values given in Table 93)} \qquad (9.25)$$

where:

F is the basic load for a split-ring or shear-plate connector unit under single shear loading (Tables 88 and 92 respectively).

It is to be noted that if load is applied at an angle α to the grain of timber, F is obtained from:

$$F = \frac{F_{/\!/}F_{\perp}}{F_{/\!/}\sin^2\alpha + F_{\perp}\cos^2\alpha}$$

$K_{62} - K_{66}$ are the modification factors for duration of loading

where:

$K_{62/66} = 1.00$, for long-term loads
       $= 1.25$, for medium-term loads
       $= 1.50$, for short and very short-term loads.

$K_{63} = K_{67}$ are the modification factors for moisture content

where:

$K_{63/67} = 1.00$, for a joint in service classes 1 and 2
       $= 0.70$, for a joint made in timber of service class 3.

$K_{64} = K_{68}$ are the modification factors for end distance

where:

$K_{64/68} = 1.00$, provided that the standard edge distance, end distance and spacings given in Tables 85–87 are used, otherwise K_{64} and K_{68} are equal to the lesser of K_s, K_C and K_D as given in Tables 87, 89 and 90 of the code.

$K_{65} = K_{69}$ are the modification factors for the number of connector units of the same size in each line for an axially loaded member, loaded parallel to the line of connectors

where:

$$K_{65/69} = 1 - \frac{3(n-1)}{100} \qquad \text{for } n < 10$$

$$= 0.7 \qquad \text{for } n \geq 10$$

$$= 1.0 \qquad \text{for all other loading cases}$$

where n is the number of connector units in each line.

9.11.3 *Metal-plate connectors*

These are light-gauge metal-plates with integral projections (teeth) punched out in one direction and bent perpendicular to the base of the plate, and are

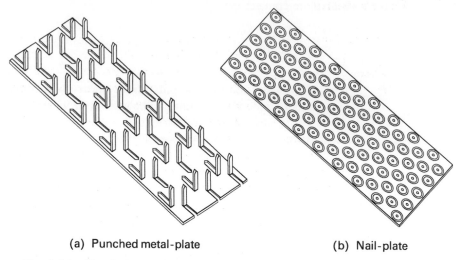

(a) Punched metal-plate   (b) Nail-plate

Fig. 9.19   Metal-plate connectors.

referred to as *punched metal-plates*. Metal-plates with pre-formed holes that depend on nails for load transfer may also fall into this category (and are referred to as *nail-plates*), see Fig. 9.19. Both types are generally used to join two or more sections of timber of the same thickness in one plane.

Punched metal-plates and nail-plates are generally manufactured from pre-galvanised mild steel strips or stainless steel strips. Their thickness ranges from 0.9 mm to 2.5 mm for punched metal-plates and around 1.0 mm for nail-plates. Manufacturers of nail-plates recommend the use of improved nails, such as square twisted nails, which should be a tight driven fit in the plate holes.

Punched metal-plate fasteners are suited to factory prefabrication and are able to transfer member forces via small connection areas. They are widely used to form truss connections as well as joints in many other plane-frame timber structures. Loads are transferred in a punched metal-plate from the timber member into the plate teeth, then from the teeth into the steel plate and across the joint interface, then back down into the teeth in the other member. Joints are designed and constructed with pairs of plates on their opposite sides.[1,3]

Many different forms of punched metal-plate fastener have been developed, involving a variety of nail patterns, nail sizes and lengths and nail shapes. Most of these are manufactured and used by individual manufacturers in their prefabricated timber products, for example, roof trusses. As such, their strength and slip characteristics are not available in the literature, but they still have to satisfy the recommendations of the related British Standards, in particular BS 5268 : Part 3 : 1998 *Code of practice for trussed rafter roofs*.

9.12   Glued joints

Another form of structural jointing system in timber is gluing, with a history going back many centuries. If gluing is used correctly it can offer the advantages of strong, rigid and durable joints with a neat appearance. Since the beginning of this century modern synthetic adhesives have been used successfully in many structural applications, including glued laminated constructions and finger-jointing systems.

Nowadays there are several types and forms of adhesives available that can be applied to timber and wood based products; but those for use in the assembly of structural components should comply with the requirements of BS EN 301 : 1992 *Adhesives, phenolic and aminoplastic, for load-bearing timber structures: classification and performance requirements*, and/ or BS 1204: 1993 *Specification for type MR phenolic and aminoplastic synthetic resin adhesives for wood.*

In glued joints, the glue should always be stronger than the surrounding timber (i.e. if it is tested to destruction, failure should occur in timber and not in the thickness of the glueline). A glued joint has a much lower strength in tension than in shear and for this reason joints should be designed so that any tension effect on the glueline is prevented. BS 5268 : Part 2, Clause 6.10 recommends that eccentric glued lap joints which induce a tensile component of stress perpendicular to the plane of the glueline should not be permitted. Therefore overlapped glued joints should always have symmetrical member and loading arrangements to induce pure shear at gluelines, see Fig. 9.20.

9.12.1   *Durability classification*

The performance of glues, particularly in relation to durability, is an important criterion for an appropriate selection of a glue type and specification by the designer. BS 5268 : Part 2, Clause 6.10.1.2 recommends that the adhesive used should be appropriate to environmental conditions in which the joint will be used. Table 94 of the code provides details of four exposure categories, the permitted adhesive types and the British Standard classifications and references.

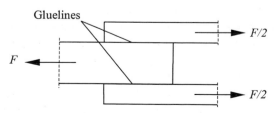

Fig. 9.20   An overlapped glued joint.

9.12.2   *Design considerations for glued joints*

As mentioned earlier, if a glued timber joint complies with the required recommendations, the glueline(s) will be subjected to pure shear and will be stronger than the surrounding timber. Therefore the design capacity of a glued connection can be calculated as the permissible shear stress on the interfaces of the timber or wood panel products being glued. Service class, load-duration and load-sharing modification factors (K_2, K_3 and K_8) apply because stresses in the connection are those of timber stresses.

The glued joint must be held in close contact during curing, and the pressure is best applied by clamps which are removed after curing, or by nails or staples which are left in the final assembly. In the event where bonding pressure is generated by nails or staples, the code recommends that the permissible shear stress be multiplied by the nail/glue modification factor, K_{70}, which has the value 0.9, i.e. $K_{70} = 0.9$.

In the event of a glued joint being designed so that one face is loaded at an angle α to the direction of the grain, the permissible shear stress for the glueline should be calculated using the following equation:

$$\tau_\alpha = \tau_{adm,/\!/}(1 - 0.67\sin\alpha) \tag{9.26}$$

where

α is the angle between the direction of the load and the grain of the piece

$\tau_{adm,/\!/}$ is the permissible shear stress parallel to the grain for the timber and may be determined from:

$$\tau_{adm,/\!/} = \tau_{g,/\!/} \times K_2K_3K_8K_{70} \tag{9.27}$$

The permissible shear stress for a joint made with different timbers and or wood based panel products (i.e. plywood or particleboard) should be taken as the permissible shear stress for the timber or the appropriate rolling shear stress for the wood based product, whichever has the lower value.

9.13   References

1. *Timber Engineering – STEP 1, Section C* (1995) Centrum Hout, The Netherlands.

2. TRADA (1995) Dowel-type fasteners for structural timber. *Wood Information*, Section 2/3, Sheet 52. Timber Research and Development Association (TRADA), High Wycombe.

3. TRADA (1996) Connectors and metal plate fasteners for structural timber. *Wood Information*, Section 2/3, Sheet 53. TRADA, High Wycombe.

9.14   Design examples

Example 9.1   Design of a simple nailed connection

The nailed joint shown in Fig. 9.21 is subjected to medium-term tensile loading under service class 2. The joint comprises six 3.00 mm diameter × 60 mm long round wire nails acting in single shear in C18 timber. Determine the load-carrying capacity of the connection.

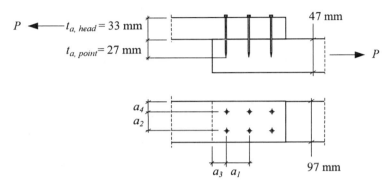

Fig. 9.21   Nailed joint details (Example 9.1).

Definitions

Force, kN	$N := newton$
Length, m	$kN := 10^3 \cdot N$
Cross-sectional dimensions, mm	Direction parallel to grain, $//$
Stress, Nmm^{-2}	Direction perpendicular to grain, pp

1.   *Geometrical details*

Nail diameter, d	$d := 3.00 \cdot mm$
Nail length, l_{nail}	$l_{nail} := 60 \cdot mm$
No. of nails, n_{nail}	$n_{nail} := 6$
Head side penetration, $t_{a,head}$	$t_{a.head} := 33 \cdot mm$
Pointside penetration, $t_{a,point}$	$t_{a.point} := l_{nail} - t_{a.head}$
	$t_{a.point} = 27 \circ mm$

2.   *Basic values*

BS 5268 : Part 2, Table 54
Strength class, C18 timber

Standard penetration, $t_{standard}$	$t_{standard} := 36 \cdot mm$
Basic single shear lateral load, F_{basic}	$F_{basic} := 308 \cdot N$

3. *Modification factors*

BS 5268 : Part 2, Clause 6.4

Number of shear planes, n_{shear} $n_{shear} := 1$

No pre-drilling $K_{pre.drill} := 1$

Nail bearing is the lesser of $K_{nail.bearing.1} := \dfrac{t_{a.head}}{t_{standard}}$

$K_{nail.bearing.1} = 0.92$

and $K_{nail.bearing.2} := \dfrac{t_{a.point}}{t_{standard}}$

$K_{nail.bearing.2} = 0.75$

Mathcad command to select $K_{nail.bearing} := [K_{nail.bearing.1} \; K_{nail.bearing.2}]$
the minimum value $\min(K_{nail.bearing}) = 0.75 \quad > 0.66$ (Clause 6.4.4.1)

okay

Thus $K_{nail.bearing} := \min(K_{nail.bearing})$

Side grain nailing, K_{43} does $K_{43} := 1$
  not apply

Ordinary nails, K_{44} does $K_{44} := 1$
  not apply

Timber-to-timber nailing, K_{46} $K_{46} := 1$
  does not apply

Load duration, K_{48}, medium-term $K_{48} := 1.12$

Moisture content, K_{49}, service $K_{49} := 1$
  class 2

No. of nails, <10, in one line, K_{50} $K_{50} := 1$

4. *Permissible load*

Permissible shear load per nail $F_{adm} := F_{basic} \cdot n_{shear} \cdot K_{pre.drill} \cdot K_{nail.bearing}$
$\cdot K_{43} \cdot K_{44} \cdot K_{46} \cdot K_{48} \cdot K_{49} \cdot K_{50}$

$F_{adm} = 258.72 \circ \text{N}$

Load capacity of connection, $P_{capacity} := F_{adm} \cdot n_{nail}$
$P_{capacity}$ $P_{capacity} = 1.55 \circ \text{kN}$

5. *Minimum nail spacings*

BS 5268 : Part 2, Table 53

For softwoods other than Douglas fir, nail spacing should be multiplied by 0.8 but the edge distance $\geq 5d$.

End distance // to grain, a_3 $a_3 := 0.8 \cdot 20 \cdot d$
$a_3 = 48 \circ \text{mm}$

Edge distance perpendicular $a_4 := 5 \cdot d$
  to grain, a_4 $a_4 = 15 \circ \text{mm}$

Distance between lines of nails perpendicular to grain, a_2
$$a_2 := 0.8 \cdot 10 \cdot d$$
$$a_2 = 24 \circ \text{mm}$$

Distance between adjacent nails in any one line // to grain, a_1
$$a_1 := 0.8 \cdot 20 \cdot d$$
$$a_1 = 48 \circ \text{mm}$$

6. *Joint slip*

BS 5268 : Part 2, Clause 6.2 and Table 52
C18 timber
Characteristic density (Table 7) $\rho := 320 \cdot \text{kg} \cdot \text{m}^{-3}$

Slip modulus:
Since the equation for slip modulus is an empirical one (i.e. with hidden units in the coefficient part), for Mathcad to produce correct units, multiply it by the units shown inside the parentheses; or alternatively, use the equation without units for ρ and d.

Thus, slip modulus, K_{ser}
$$K_{ser} := \tfrac{1}{25} \cdot \rho^{1.5} \cdot d^{0.8} \cdot (\text{kg}^{-1.5} \cdot \text{m}^{2.7} \cdot \text{N} \cdot 10^{5.4})$$
$$K_{ser} = 551.42 \circ \text{N} \cdot \text{mm}^{-1}$$

Slip per nail, u, under permissible load
$$u := \frac{F_{adm}}{K_{ser}}$$
$$u = 0.47 \circ \text{mm}$$

Therefore the load capacity of the joint is 1.55 kN with a joint slip of 0.47 mm

Example 9.2 Design of a timber-to-timber connection with nailing into the end grain

The nailed connection shown in Fig. 9.22 is subjected to short-term loading under service class 2 conditions. The joint comprises two 3.75 mm diameter × 65 mm long round wire nails acting in single shear in C22 timber. Determine the load capacity of this connection.

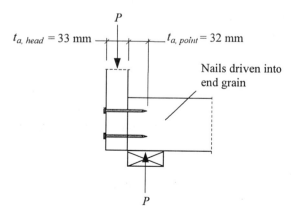

Fig. 9.22 Nailed joint details (Example 9.2).

Definitions

Force, kN	$N := newton$
Length, m	$kN := 10^3 \cdot N$
Cross-sectional dimensions, mm	Direction parallel to grain, $//$
Stress, Nmm^{-2}	Direction perpendicular to grain, pp

1. Geometrical details

Nail diameter, d	$d := 3.75 \cdot mm$
Nail length, l_{nail}	$l_{nail} := 65 \cdot mm$
No. of nails, n_{nail}	$n_{nail} := 2$
Head side penetration, $t_{a,head}$	$t_{a.head} := 33 \cdot mm$
Pointside penetration, $t_{a,point}$	$t_{a.point} := l_{nail} - t_{a.head}$
	$t_{a.point} = 32 \circ mm$

2. Basic values

BS 5268 : Part 2, Table 54
Strength class, C22 timber

Standard penetration, $t_{standard}$	$t_{standard} := 45 \cdot mm$
Basic single shear lateral load, F_{basic}	$F_{basic} := 444 \cdot N$

3. Modification factors

BS 5268 : Part 2, Clause 4.6

Number of shear planes, n_{shear}	$n_{shear} := 1$
No pre-drilling	$K_{pre.drill} := 1$
Nail bearing is the lesser of	$K_{nail.bearing.1} := \dfrac{t_{a.head}}{t_{standard}}$
	$K_{nail.bearing.1} = 0.73$
and	$K_{nail.bearing.2} := \dfrac{t_{a.point}}{t_{standard}}$
	$K_{nail.bearing.2} = 0.71$
Mathcad command to select the minimum value	$K_{nail.bearing} := [K_{nail.bearing.1} \ K_{nail.bearing.2}]$
	$\min(K_{nail.bearing}) = 0.71 \quad > 0.66 \quad okay$
Thus	$K_{nail.bearing} := \min(K_{nail.bearing})$
End grain nailing, K_{43} does not apply	$K_{43} := 0.7$
Ordinary nails, K_{44} does not apply	$K_{44} := 1$
Timber-to-timber nailing, K_{46} does not apply	$K_{46} := 1$

Load duration, K_{48}, short-term $K_{48} := 1.25$

Moisture content, K_{49}, service $K_{49} := 1$
  class 2

No. of nails, <10, in one line, K_{50} $K_{50} := 1$

4. Permissible load

BS 5268 : Part 2, Clause 4.6.9

Permissible shear load per nail $F_{adm} := F_{basic} \cdot n_{shear} \cdot K_{pre.drill} \cdot K_{nail.bearing}$
$$\cdot K_{43} \cdot K_{44} \cdot K_{46} \cdot K_{48} \cdot K_{49} \cdot K_{50}$$
$$F_{adm} = 276.27 \circ \text{N}$$

Load capacity of connection, $P_{capacity} := F_{adm} \cdot n_{nail}$
  $P_{capacity}$ $P_{capacity} = 0.55 \circ \text{kN}$

5. Joint slip

BS 5268 : Part 2, Clause 6.2 and Table 52
C22 timber

Characteristic density (Table 7) $\rho := 340 \cdot \text{kg} \cdot \text{m}^{-3}$

Slip modulus:

Since the equation for slip modulus is an empirical one (i.e. with hidden units in the coefficient part), for Mathcad to produce correct units, multiply it by the units shown inside the parentheses; or alternatively, use the equation without units for ρ and d.

Thus, slip modulus, K_{ser} $K_{ser} := \frac{1}{25} \cdot \rho^{1.5} \cdot d^{0.8} \cdot (\text{kg}^{-1.5} \cdot \text{m}^{2.7} \cdot \text{N} \cdot 10^{5.4})$
$$K_{ser} = 721.94 \circ \text{N} \cdot \text{mm}^{-1}$$

Slip per nail, u, under $u := \dfrac{F_{adm}}{K_{ser}}$
  permissible load
$$u = 0.38 \circ \text{mm}$$

Therefore the load capacity of the joint is 0.55 kN with a joint slip of 0.38 mm

Example 9.3   Design of a timber-to-timber nailed connection in double shear

The nailed joint shown in Fig. 9.23 is subjected to long-term tensile loading under service class 1 conditions. The joint comprises six 3.35 mm diameter × 100 mm long helical-threaded shank nails acting in double shear in C18 timber. Determine the load capacity of this connection.

Definitions

Force, kN   N := newton

Length, m $kN := 10^3 \cdot \text{N}$

Cross-sectional dimensions, mm   Direction parallel to grain, //

Stress, Nmm^{-2} Direction perpendicular to grain, *pp*

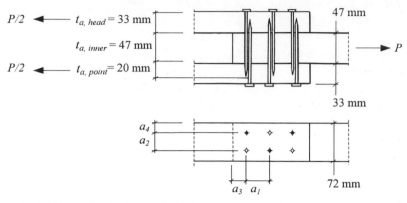

Fig. 9.23   Naileld joint details (Example 9.3).

1. *Geometrical details*

Nail diameter, d	$d := 3.35 \cdot \text{mm}$
Nail length, l_{nail}	$l_{nail} := 100 \cdot \text{mm}$
No. of nails, n_{nail}	$n_{nail} := 6$
Head side penetration, $t_{a,head}$	$t_{a.head} := 33 \cdot \text{mm}$
Inner member, $t_{a,inner}$	$t_{a.inner} := 47 \cdot \text{mm}$
Pointside penetration, $t_{a,point}$	$t_{a.point} := l_{nail} - t_{a.head} - t_{a.inner}$
	$t_{a.point} = 20 \circ \text{mm}$

2. *Basic values*

BS 5268 : Part 2, Table 54
Strength class, C18 timber

Standard penetration, $t_{standard}$	$t_{standard} := 40 \cdot \text{mm}$
Basic single shear lateral load, F_{basic}	$F_{basic} := 369 \cdot \text{N}$

3. *Modification factors*

BS 5268 : Part 2, Clause 4.6

Number of shear planes, n_{shear}	$n_{shear} := 2$
No pre-drilling	$K_{pre.drill} := 1$

Nail bearing is the lesser of

$$K_{nail.bearing.1} := \frac{t_{a.head}}{t_{standard}}$$

$$K_{nail.bearing.1} = 0.83$$

and

$$K_{nail.bearing.2} := \frac{t_{a.inner}}{0.85 \cdot t_{standard}}$$

$$K_{nail.bearing.2} = 1.38$$

and

$$K_{nail.bearing.3} := \frac{t_{a.point}}{t_{standard}}$$

$$K_{nail.bearing.3} = 0.5$$

Mathcad command to select the minimum value

$$K_{nail.bearing} := [K_{nail.bearing.1} \ K_{nail.bearing.2} \ K_{nail.bearing.3}]$$

$$\min(K_{nail.bearing}) = 0.5 \quad \text{(Clause 6.4.4.1)} \quad \text{okay}$$

Thus

$$K_{nail.bearing} := \min(K_{nail.bearing})$$

Side grain nailing, K_{43} does not apply

$$K_{43} := 1$$

Improved nails, K_{44} does not apply

$$K_{44} := 1.2$$

Timber-to-timber nailing, K_{46} does not apply

$$K_{46} := 1$$

Load duration, K_{48}, long-term

$$K_{48} := 1$$

Moisture content, K_{49}, service class 2

$$K_{49} := 1$$

No. of nails, <10, in one line, K_{50}

$$K_{50} := 1$$

4. Permissible load

BS 5268 : Part 2, Clause 4.6.9

Permissible shear load per nail

$$F_{adm} := F_{basic} \cdot n_{shear} \cdot K_{pre.drill} \cdot K_{nail.bearing}$$
$$\cdot K_{43} \cdot K_{44} \cdot K_{46} \cdot K_{48} \ K_{49} \ K_{50}$$

$$F_{adm} = 442.8 \circ \text{N}$$

Load capacity of connection, $P_{capacity}$

$$P_{capacity} := F_{adm} \cdot n_{nail}$$
$$P_{capacity} = 2.66 \circ \text{kN}$$

5. Minimum nail spacings

BS 5268 : Part 2, Table 53

For softwoods other than Douglas fir, nail spacing should be multiplied by 0.8 but the edge distance $\geq 5d$.

End distance // to grain, a_3

$$a_3 := 0.8 \cdot 20 \cdot d$$
$$a_3 = 53.6 \circ \text{mm}$$

Edge distance perpendicular to grain, a_4

$$a_4 := 5 \cdot d$$
$$a_4 = 16.75 \circ \text{mm}$$

Distance between lines of nails perpendicular to grain, a_2

$$a_2 := 0.8 \cdot 10 \cdot d$$
$$a_2 = 26.8 \circ \text{mm}$$

Distance between adjacent nails in any one line // to grain, a_1

$$a_1 := 0.8 \cdot 20 \cdot d$$
$$a_1 = 53.6 \circ \text{mm}$$

6. Joint slip

BS 5268 : Part 2, Clause 6.2 and Table 52

C18 timber

Characteristic density (Table 7)

$$\rho := 320 \cdot \text{kg} \cdot \text{m}^{-3}$$

Slip modulus:

Since the equation for slip modulus is an empirical one (i.e. with hidden units in the coefficient part), for Mathcad to produce correct units, multiply it by the units shown inside the parentheses; or alternatively, use the equation without units for ρ and d.

Thus, slip modulus, K_{ser}

$$K_{ser} := \tfrac{1}{25} \cdot \rho^{1.5} \cdot d^{0.8} \cdot (\text{kg}^{-1.5} \cdot \text{m}^{2.7} \cdot \text{N} \cdot 10^{5.4})$$
$$K_{ser} = 602.31 \circ \text{N} \cdot \text{mm}^{-1}$$

Slip per nail, u, under permissible load

$$u := \frac{F_{adm}}{K_{ser}}$$
$$u = 0.74 \circ \text{mm}$$

Therefore the load capacity of the joint is 2.66 kN with a joint slip of 0.74 mm

Example 9.4   Design of a nailed plywood gusset eave joint

Design of an eave joint of a roof truss is required. The joint is to comprise a Finnish conifer plywood (9 mm thick, 7-ply) gusset with 3.00 mm diameter × 50 mm long round wire nails acting in single shear. The rafter and ceiling tie are both in C18 timber under service class 2 conditions and are subjected to medium-term loading, as shown in Fig. 9.24(a).

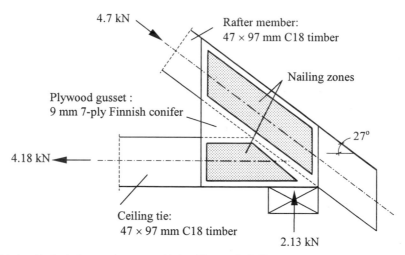

Fig. 9.24(a)   Nailed plywood gusseted joint (Example 9.4).

Definitions

Force, kN $\text{N} := \text{newton}$
Length, m $\text{kN} := 10^3 \cdot \text{N}$
Cross-sectional dimensions, mm   Direction parallel to grain, //
Stress, Nmm^{-2} Direction perpendicular to grain, pp

1. Geometrical details

Nail diameter, d $d := 3.00 \cdot mm$

Nail length, l_{nail} $l_{nail} := 50 \cdot mm$

No. of nails, n_{nail} Required

Head side penetration, $t_{a,head}$ $t_{a.head} := 9 \cdot mm$

Pointside penetration, $t_{a,point}$ $t_{a.point} := l_{nail} - t_{a.head}$

$t_{a.point} = 41 \circ mm$

2. Basic values

BS 5268 : Part 2, Table 56
Plywood group I and C18 timber
Basic single shear lateral $F_{basic} := 267 \cdot N$
    load, F_{basic}

3. Modification factors

BS 5268 : Part 2, Clause 4.6

Number of shear planes, n_{shear} $n_{shear} := 1$

No pre-drilling $K_{pre.drill} := 1$

Side grain nailing, K_{43} does $K_{43} := 1$
    not apply

Ordinary nails, K_{44} does $K_{44} := 1$
    not apply

Plywood-to-timber, K_{46} does $K_{46} := 1$
    not apply

Load duration, K_{48}, medium-term $K_{48} := 1.12$

Moisture content, K_{49}, service $K_{49} := 1$
    class 2

No. of nails, assume <10, in one $K_{50} := 1$
    line, K_{50}

4. Permissible load

BS 5268 : Part 2, Clause 4.6.9

Permissible shear lateral load $F_{adm} := F_{basic} \cdot n_{shear} \cdot K_{pre.drill}$
    per nail $\cdot K_{43} \cdot K_{44} \cdot K_{46} \cdot K_{48} \cdot K_{49} \cdot K_{50}$

$F_{adm} = 299.04 \circ N$

5. Minimum nail spacings

BS 5268 : Part 2, Table 53

End distance // to grain, a_3 $a_3 := 14 \cdot d$

$a_3 = 42 \circ mm$

Edge distance perpendicular to grain, a_4	$a_4 := 5 \cdot d$
	$a_4 = 150 \, \text{mm}$
Distance between lines of nails perpendicular to grain, a_2	$a_2 := 7 \cdot d$
	$a_2 = 210 \, \text{mm}$
Distance between adjacent nails in any one line // to grain, a_1	$a_1 := 14 \cdot d$
	$a_1 = 420 \, \text{mm}$

6. Rafter connection

Load in rafter, P_{rafter}	$P_{rafter} := 4.7 \cdot \text{kN}$
Number of nails required, n_{rafter}	$n_{rafter} := \dfrac{P_{rafter}}{F_{adm}}$
	$n_{rafter} = 15.72$
Mathcad command	$n_{rafter} := \text{ceil}(n_{rafter})$
	Adopt: $n_{rafter} = 16$

Using 2 rows of 8 nails each will give a minimum required gusset length parallel to rafter of, L_{rafter}

$$L_{rafter} := 2 \cdot a_3 + 7 \cdot a_1$$
$$L_{rafter} = 378 \, \text{mm}$$

7. Ceiling tie connection

Load in ceiling tie, P_{tie}	$P_{tie} := 4.18 \cdot \text{kN}$
Number of nails required, n_{rafter}	$n_{rafter} := \dfrac{P_{tie}}{F_{adm}}$
	$n_{rafter} = 13.98$
Mathcad command	$n_{rafter} := \text{ceil}(n_{rafter})$
	Adopt: $n_{rafter} = 14$

Using 3 rows of nails, 4 on top, 4 along the centre line and 5 at the bottom row, will give a minimum required length for gusset along the bottom nailing row of the ceiling tie of, L_{tie}

$$L_{tie} := 2 \cdot a_3 + 5 \cdot a_1$$
$$L_{tie} = 294 \, \text{mm}$$

Thus, a suitable gusset length along the bottom edge of the ceiling tie should be

$$L_{rafter} \cdot \cos\left(27 \cdot \frac{\pi}{180}\right) = 336.8 \, \text{mm} > 294 \, \text{mm}$$

Try a gusset of 350 mm in length along the bottom edge of the ceiling tie.

8. Design of the plywood gusset

A free-body diagram of the non-concurrent forces acting on the joint, with reference to the centre line of the members, is shown in Fig. 9.24(b). Since the three forces are not

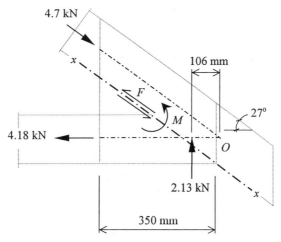

Fig. 9.24(b)   Free-body diagram.

coincident the free-body is not in full equilibrium (i.e. the sum of the moments of the forces (say about point O) is not equal to zero). Thus, the applied non-concurrent force system will induce a moment of M, at the joint, which should be resisted by the plywood gusset plate. In addition, the plywood gusset should also be able to resist a shearing force of $F - 4.7\,\mathrm{kN}$ acting along an axis parallel to the rafter and passing through the interface between the ceiling tie and the rafter (i.e. along x–x).

Therefore, the balancing moment, M, is	$M := (2.13 \cdot \mathrm{kN}) \cdot (106 \cdot \mathrm{mm})$ $M = 0.23 \circ \mathrm{kN} \cdot \mathrm{m}$
Plywood thickness, t	$t := 9 \cdot \mathrm{mm}$
Length of the gusset along the rafter, h	$h := (350 \cdot \mathrm{mm}) \cdot \dfrac{1}{\cos\left(27 \cdot \dfrac{\pi}{180}\right)}$ $h = 392.81 \circ \mathrm{mm}$
Section modulus, z	$z := \dfrac{t \cdot h^2}{6}$ $z = 2.31 \times 10^5 \circ \mathrm{mm}^3$
Stress due to bending moment	$\sigma_m := \dfrac{M}{z}$ $\sigma_m = 0.98 \circ \mathrm{N} \cdot \mathrm{mm}^{-2}$
Shearing stress along x–x	$\sigma_s := \dfrac{4.7 \cdot \mathrm{kN}}{t \cdot h}$ $\sigma_s = 1.33 \circ \mathrm{N} \cdot \mathrm{mm}^{-2}$
Total stress acting along x–x	$\sigma_{total} := \sigma_m + \sigma_s$ $\sigma_{total} = 2.3 \circ \mathrm{N} \cdot \mathrm{mm}^{-2}$

BS 5268 : Part 2, Table 41

Grade panel shear $\qquad\qquad \tau_g := 3.74 \cdot N \cdot mm^{-2}$

Modification factor for plywood $\quad K_{36} := 1.33$
grade stresses for medium-term
(K_{36}, Table 33)

Permissible panel shear in $\qquad \tau_{adm} := \tau_g \cdot K_{36}$
plywood $\qquad\qquad\qquad\qquad \tau_{adm} = 4.97 \circ N \cdot mm^{-2}$

Plywood size is satisfactory

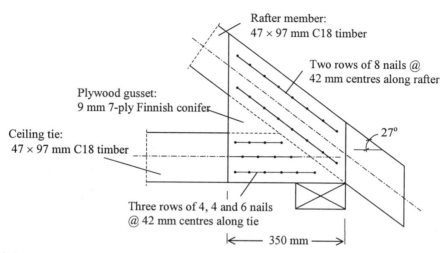

Rafter member:
47 × 97 mm C18 timber

Two rows of 8 nails @
42 mm centres along rafter

Plywood gusset:
9 mm 7-ply Finnish conifer

Ceiling tie:
47 × 97 mm C18 timber

27°

Three rows of 4, 4 and 6 nails
@ 42 mm centres along tie

350 mm

Fig. 9.24(c)   Designed joint details.

Example 9.5   *Design of a simple two-member timber-to-timber bolted connection*

The bolted connection shown in Fig. 9.25 is subjected to medium-term tensile loading under service class 2 conditions. The joint comprises a single M12 bolt acting in single shear with C18 timber sections. Determine the load capacity of this connection.

Definitions

Force, kN $\qquad\qquad\qquad\qquad$ N := newton

Length, m $\qquad\qquad\qquad\qquad$ kN := $10^3 \cdot$ N

Cross-sectional dimensions, mm $\quad$ Direction parallel to grain, //

Stress, Nmm^{-2} $\qquad\qquad\qquad$ Direction perpendicular to grain, *pp*

1.   *Geometrical details*

Bolt diameter, d $\qquad\qquad\qquad d := 12 \cdot mm$

No. of bolts, n_{bolt} $\qquad\qquad\qquad n_{bolt} := 1$

Thinner member thickness $\qquad\quad t_{thinner} := 33 \cdot mm$

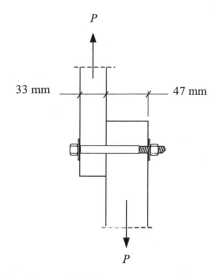

33 mm   47 mm

Fig. 9.25   Bolted connection (Example 9.5).

2. *Basic values*

BS 5268 : Part 2, Table 64
Strength class, C18 timber, load parallel to grain
Basic single shear lateral load $F_{basic} := 1.22 \cdot kN$
   under medium-term loading,
   by interpolation, F_{basic}

3. *Modification factors*

BS 5268 : Part 2, Clause 6.6
Timber-to-timber connection $K_{46} := 1$
   K_{46} does not apply
Moisture content, K_{56}, service $K_{56} := 1$
   class 2
No. of bolts in one line, $K_{57} := 1$
   $n = 1$, K_{57}

4. *Permissible load*

BS 5268 : Part 2, Clause 6.6.6
Permissible shear lateral load $F_{adm} := F_{basic} \cdot K_{46} \cdot K_{56} \cdot K_{57}$
   per bolt $F_{adm} = 1.22 \circ kN$
Load capacity of the connection, $P_{capacity} := F_{adm} \cdot n_{bolt}$
   $P_{capacity}$ $P_{capacity} = 1.22 \circ kN$

5.   Minimum bolt end and edge distances

BS 5268 : Part 2, Table 75

End distance loaded	$7 \cdot d = 84 \circ$ mm
End distance unloaded	$4 \cdot d = 48 \circ$ mm not applicable
Edge distance loaded	$4 \cdot d = 48 \circ$ mm not applicable
Edge distance unloaded	$1.5 \cdot d = 18 \circ$ mm

6.   Joint slip

BS 5268 : Part 2, Clause 6.2 and Table 52
Strength class, C18 timber
Characteristic density (Table 7) $\rho := 320 \cdot kg \cdot m^{-3}$

Slip modulus:
Since the equation for slip modulus is an empirical one (i.e. with hidden units in the coefficient part), for Mathcad to produce correct units, muliply it by the units shown inside the parentheses; or alternatively, use the equation without units for ρ and d.

Thus, slip modulus, K_{ser}
$$K_{ser} := \tfrac{1}{20} \cdot \rho^{1.5} \cdot d \cdot (kg^{-1.5} \cdot m^{2.5} \cdot N \cdot 10^6)$$
$$K_{ser} = 3.43 \times 10^3 \circ N \cdot mm^{-1}$$

Slip per bolt under permissible
load, u
$$u := \frac{F_{adm}}{K_{ser}}$$
$$u = 0.36 \circ mm$$

Allowing 1.00 mm additional slip
due to oversize bolt hole
$$u_{joint} := u + 1.00 \cdot mm$$
$$u_{joint} = 1.36 \circ mm$$

Therefore the load capacity of the joint is 1.22 kN with a joint slip of 1.36 mm

Example 9.6   Design of a three-member timber-to-timber bolted connection

The bolted connection shown in Fig. 9.26 is subjected to a long-term loading, as indicated, under service class 2 conditions. The joint comprises a single M16 bolt acting in double shear with C22 timber sections. Check the adequacy of the connection.

Definitions

Force, kN	$N := $ newton
Length, m	$kN := 10^3 \cdot N$
Cross-sectional dimensions, mm	Direction parallel to grain, //
Stress, Nmm^{-2}	Direction perpendicular to grain, *pp*

1.   Geometrical details

Bolt diameter, d	$d := 16 \cdot mm$
No. of bolts, n_{bolt}	$n_{bolt} := 1$

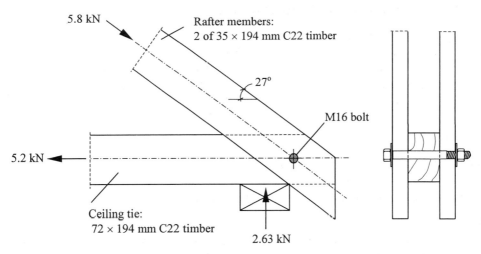

Fig. 9.26 Bolted connection (Example 9.6).

Side member thickness	$t_{outer} := 35 \cdot mm$
Inner member thickness	$t_{inner} := 72 \cdot mm$
Check adequacy of the section	$t_{inner} = 72 \circ mm > 2 \cdot t_{outer} = 70 \circ mm$ okay
Distance from loaded edge to furthest fastener	$h_e := \dfrac{194 \cdot mm}{2}$
	$h_e = 97 \circ mm$

2. *Basic values*

BS 5268 : Part 2, Clause 6.6.4 and Table 70
The support reaction acting on the ceiling tie makes the load in the rafter act parallel to grain, and in the ceiling tie to act at 27° to grain direction. Therefore Hankinson's equation should be used to determine the appropriate basic shear load values.

Strength class, C22 timber (Table 70)
(1) For 35 mm outer members (rafters), long-term load parallel to the grain for a M16 bolt $F_{rafter.\parallel} := 3.09 \cdot kN$

(2) For 72 mm inner member (ceiling tie) corresponding to 72/2 = 36 mm outer members for a M16 bolt:

long-term load parallel to grain, by interpolation $F_{tie.\parallel} := 3.18 \cdot kN$

long-term load perpendicular to grain, by interpolation $F_{tie.pp} := 2.64 \cdot kN$

For load at angle α where

$$\alpha := 27 \cdot \frac{\pi}{180}$$

$$F_{tie} := \frac{F_{tie.\parallel} \cdot F_{tie.pp}}{F_{tie.\parallel} \cdot \sin(\alpha)^2 + F_{tie.pp} \cdot \cos(\alpha)^2}$$

$$F_{tie} = 3.05 \circ kN$$

(3)   Grade shear stress for C22 timber (Table 7)

$$\tau_{g.\parallel} := 0.71 \cdot N \cdot mm^{-2}$$

3.   *Modification factors*

BS 5268 : Part 2, Clause 6.6

Timber-to-timber connection, K_{46} does not apply

$$K_{46} := 1$$

Moisture content, K_{56}, service class 2

$$K_{56} := 1$$

No. of bolts in one line, $n = 1$, K_{57}

$$K_{57} := 1$$

Service class 2 (K_2, Table 13)

$$K_2 := 1$$

Load duration, long-term (K_3, Table 14)

$$K_3 := 1$$

Load sharing (K_8, Clause 2.9)

$$K_8 := 1$$

4.   *Permissible loads*

(1)   For rafter:

Permissible load per shear plane

$$F_{adm.rafter} := F_{rafter.\parallel} \cdot K_{46} \cdot K_{56} \cdot K_{57}$$
$$F_{adm.rafter} = 3.09 \circ kN$$

actual load (per shear plane)

$$P_{rafter} := \frac{5.80 \cdot kN}{2}$$

$$P_{rafter} = 2.9 \circ kN$$
Satisfactory

(2)   For ceiling tie:

permissible load per shear plane

$$F_{adm.tie} := F_{tie} \cdot K_{46} \cdot K_{56} \cdot K_{57}$$
$$F_{adm.tie} = 3.05 \circ kN$$

actual load (per shear plane)

$$P_{rafter} := \frac{5.20 \cdot kN}{2}$$

$$P_{rafter} = 2.6 \circ kN$$
Satisfactory

(3)   Shear stress in jointed timber
BS 5268 : Part 2, Clause 6.1 and Figure 6

Permissible shear stress

$$\tau_{adm} := \tau_{g.\parallel} \cdot K_2 \cdot K_3 \cdot K_8$$
$$\tau_{adm} = 0.71 \circ N \cdot mm^{-2}$$

Permissible shear force

$$V_{adm} := \tfrac{2}{3} \cdot \tau_{adm} \cdot h_e \cdot t_{inner}$$
$$V_{adm} = 3.31 \circ kN$$
$$> 2.63 \, kN \quad \text{Satisfactory}$$

5. *Minimum bolt spacings*

BS 5268 : Part 2, Table 75
(1) For rafter:

end distance loaded	$7 \cdot d = 112 \circ$ mm not applicable
end distance unloaded	$4 \cdot d = 64 \circ$ mm
edge distance loaded	$4 \cdot d = 64 \circ$ mm not applicable
edge distance unloaded	$1.5 \cdot d = 24 \circ$ mm

(2) For ceiling tie:

end distance loaded	$7 \cdot d = 112 \circ$ mm
end distance unloaded	$4 \cdot d = 64 \circ$ mm not applicable
edge distance loaded	$4 \cdot d = 64 \circ$ mm
edge distance unloaded	$1.5 \cdot d = 24 \circ$ mm

Therefore the connection with minimum edge and end distances detailed above will be adequate

Example 9.7   Design of a three-member steel-to-timber bolted moment connection

The bolted connection shown in Fig. 9.27 is subjected to a long-term eccentric load of 12.5 kN as indicated. The joint comprises two 4 mm thick steel side-plates and an inner timber member in strength class C22 joined together using eight M12 bolts under service class 2 conditions. Check the adequacy of the connection.

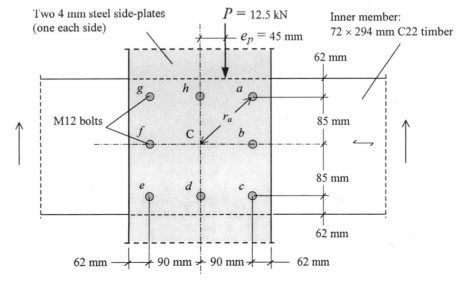

Fig. 9.27   Eccentric loaded connection (Example 9.7).

Definitions

Force, kN

Length, m

Cross-sectional dimensions, mm

Stress, Nmm^{-2}

$N :=$ newton

$kN := 10^3 \cdot N$

Direction parallel to grain, //

Direction perpendicular to grain, pp

1. Geometrical details

Bolt diameter, d

No. of bolts, n_{bolt}

Bolting pattern:

horizontal spacing, x

vertical spacing, y

Side member thickness
  (Clause 6.6.5.1)

Inner member thickness

Distance from loaded edge to
  furthest fastener

$d := 12 \cdot$ mm

$n_{bolt} := 8$

$x := 90 \cdot$ mm

$y := 85 \cdot$ mm

$t_{steel} := 4 \cdot$ mm > 2.5 mm or $0.3 \cdot d = 3.6 \circ$ mm
                                                          okay

$t_{inner} := 72 \cdot$ mm

$h_e := 294 \cdot$ mm $- 62 \cdot$ mm

$h_e = 232 \circ$ mm

2. Applied loading

Applied vertical load, P

Eccentricity, e_p

No applied horizontal load, H

Eccentricity will induce a
  moment of M and a shear
  force of V at C, the centre
  of geometry of the connection

Distance from C to each bolt, r

$P := 12.5 \cdot$ kN

$e_p := 45 \cdot$ mm

$H := 0 \cdot$ kN

$M := P \cdot e_p$ $M = 0.56 \circ$ kN $\cdot$ m

$V := P$ $Y = 12.5 \circ$ kN

$r_a := \sqrt{x^2 + y^2}$

$r_a = 123.79 \circ$ mm

$r_b := x$ $r_f := x$

$r_c := r_a$ $r_e := r_a$ $r_g := r_a$

$r_d := y$ $r_h := y$

Sum of squares of distances
  from C to each bolt, $J = \sum r^2$

$J := 4 \cdot r_a^2 + 2 \cdot r_b^2 + 2 \cdot r_d^2$

$J = 9.2 \times 10^4 \circ$ mm^2

Force acting on furthest bolt
  (i.e. bolt a) due to applied
  moment M

$F_{a.M} := \dfrac{M}{J} \cdot r_a$

$F_{a.M} = 0.76 \circ$ kN

Force acting on bolt a due to
  applied shear force V

$F_V := \dfrac{V}{n_{bolt}}$

$F_V = 1.56 \circ$ kN

Force acting on bolt a due to applied horizontal load $H = 0$

$$F_H := \frac{H}{n_{bolt}}$$

$$F_H = 0 \circ kN$$

For the rectangular fastener pattern, the total force on bolt a is given by

$$F := \left[\left(F_V + \frac{x}{\sqrt{(x^2 + y^2)}} \cdot F_{a.M} \right)^2 + \left(F_H + \frac{y}{\sqrt{(x^2 + y^2)}} \cdot F_{a.M} \right)^2 \right]^{\frac{1}{2}}$$

$$F = 2.18 \circ kN$$

This force is acting at an angle of α to grain of timber, where

$$\alpha := atan\left(\frac{x \cdot F_{a.M} + \sqrt{x^2 + y^2} \cdot F_V}{y \cdot F_{a.M} + 0} \right)$$

$$\alpha = 76.18 \circ deg$$

3.   *Basic load values*

BS 5268 : Part 2, Clause 6.6.4 and Table 70
72 mm inner member corresponding to $72/2 = 36$ mm outer members for a M12 bolt:

Long-term load // to grain, by interpolation

$$F_{//} := 2.296 \cdot kN$$

Long-term load perpendiuclar to grain, by interpolation

$$F_{pp} := 2.105 \cdot kN$$

For load acting at angle α

$$F := \frac{F_{//} \cdot F_{pp}}{F_{//} \cdot sin(\alpha)^2 + F_{pp} \cdot cos(\alpha)^2}$$

$$F = 2.12 \circ kN$$

Grade shear stress for C22 timber (Table 7)

$$\tau_{g.//} := 0.71 \cdot N \cdot mm^{-2}$$

4.   *Modification factors*

BS 5268 : Part 2, Clause 6.6

Steel-to-timber connection (K_{46}, Clause 6.6.5.1)

$$K_{46} := 1.25$$

Moisture content, K_{56}, service class 2

$$K_{56} := 1$$

No. of bolts in one line, $n = 3$, K_{57}

$$K_{57} := 1$$

Service class 2 (K_2, Table 13)

$$K_2 := 1$$

Load duration, long-term (K_3, Table 14)

$$K_3 := 1$$

Load sharing (K_8, Clause 2.9)

$$K_8 := 1$$

5. *Permissible loads*

BS 5268 : Part 2, Clause 6.6.6

Permissible load per shear plane
$$F_{adm} := F \cdot K_{46} \cdot K_{56} \cdot K_{57}$$
$$F_{adm} = 2.64 \circ kN$$

Number of shear planes
$$n_{shear} := 2$$

Permissible load for bolt
$$F_{adm.bolt} := F_{adm} \cdot n_{shear}$$
$$F_{adm.bolt} = 5.29 \circ kN$$
Satisfactory

Shear stress in jointed timber:

BS 5268 : Part 2, Clause 6.1 and Figure 6

Permissible shear stress
$$\tau_{adm} := \tau_{g.//} \cdot K_2 \cdot K_3 \cdot K_8$$
$$\tau_{adm} = 0.71 \circ N \cdot mm^{-2}$$

Permissible shear force
$$V_{adm} := \tfrac{2}{3} \cdot \tau_{adm} \cdot h_e \cdot t_{inner}$$
$$V_{adm} = 7.91 \circ kN$$

Hence the timber section can
sustain an applied shear force
at the connection of $F_{adm.V}$
$$F_{adm.V} := 2 \cdot V_{adm}$$
$$F_{adm.V} = 15.81 \circ kN$$
Satisfactory

6. *Minimum bolt spacings*

BS 5268 : Part 2, Table 75

end distance loaded	$7 \cdot d = 84 \circ mm$	not applicable
end distance unloaded	$4 \cdot d = 48 \circ mm$	not applicable
edge distance loaded	$4 \cdot d = 48 \circ mm$	$< 56\,mm$ provided okay
edge distance unloaded	$1.5 \cdot d = 18 \circ mm$	not applicable
spacing parallel to grain	$5 \cdot d = 60 \circ mm$	$< 78\,mm$ provided okay
spacing perpendicular to grain	$4 \cdot d = 48 \circ mm$	$< 73\,mm$ provided okay

Therefore the connection as detailed above is adequate

Example 9.8   Design of a glued plywood gusset eave joint

Design of an eave joint of a roof truss is required. The joint is to comprise two Finnish conifer plywood (9 mm thick, 7-ply) gussets, one on each side, glued to rafter and ceiling tie members both in C18 timber under service class 2 conditions and is subjected to medium-term loading, as shown in Fig. 9.28. Determine the contact areas required for the rafter and tie member where bonding pressure is generated by nails.

Definitions

Force, kN	$N := $ newton
Length, m	$kN := 10^3 \cdot N$
Cross-sectional dimensions, mm	Direction parallel to grain, $//$
Stress, Nmm^{-2}	Direction perpendicular to grain, *pp*

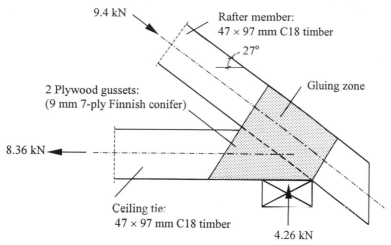

Fig. 9.28   Glued plywood gusset joint (Example 9.8).

1. *Basic shear stress values*

BS 5268 : Part 2, Clause 6.10.1.4
Because the joint is to be formed by gluing plywood gusset plates to timber, the permissible shear stress for the joint is the lesser of that for timber and rolling shear stress for the plywood.

C18 timber (Table 7)
Shear stress parallel to grain $\tau_{g.\!/\!/} := 0.67 \cdot N \cdot mm^{-2}$
Finnish conifer plywood (Table 41)
Rolling shear stress $\tau_{r.g} := 0.79 \cdot N \cdot mm^{-2}$

2. *Modification factors*

Service class 2 (K_2, Table 13) $K_2 := 1$
Medium-term loading $K_3 := 1.25$
  (K_3, Table 14)
No load sharing (K_8, Clause 2.9) $K_8 := 1$
Plywood grade stresses $K_{36} := 1.33$ medium-term loading
  (K_{36}, Table 33)
Nail/glue modification factor $K_{70} := 0.9$
  (K_{70}, Clause 6.10.1.4)

3. *Permissible stresses*

The support reaction acting on the ceiling tie makes the load in the rafter act parallel to the grain, and in the ceiling tie to act at 27° to grain direction. Thus

Angle of inclination, α $\alpha := 27 \cdot \dfrac{\pi}{180}$

Permissible shear stress for timber	$\tau_{adm.//} := \tau_{g.//} \cdot K_2 \cdot K_3 \cdot K_8 \cdot K_{70}$

for the rafter
$$\tau_{adm.rafter} := \tau_{adm.//}$$
$$\tau_{adm.rafter} = 0.75 \circ \text{N} \cdot \text{m}^{-2}$$

for the ceiling tie (Clause 6.10.1.4) $\tau_{adm.tie} := \tau_{adm.//} \cdot (1 - 0.67 \cdot \sin(\alpha))$
$$\tau_{adm.tie} = 0.52 \circ \text{N} \cdot \text{mm}^{-2}$$

permissible rolling shear stress
    for plywood
$$\tau_{adm.r} := \tau_{r.g} \cdot K_{36} \cdot K_{70}$$
$$\tau_{adm.r} = 0.95 \circ \text{N} \cdot \text{mm}^{-2}$$

Therefore the shear stress for timber is limiting for both the rafter and tie contact faces.

4. *Contact (gluing) areas*

(1) For rafter
    load in rafter, P_{rafter} $P_{rafter} := 9.4 \cdot \text{kN}$

   required contact area on
        each side, A_{rafter}
$$A_{rafter} := \frac{P_{rafter}}{2 \cdot \tau_{adm.rafter}}$$
$$A_{rafter} = 6.24 \times 10^3 \circ \text{mm}^2$$

(2) For ceiling tie
    load in tie, P_{tie} $P_{tie} := 8.36 \cdot \text{kN}$

   required contact area on
        each side, A_{tie}
$$A_{tie} := \frac{P_{tie}}{2 \cdot \tau_{adm.tie}}$$
$$A_{tie} = 7.97 \times 10^3 \circ \text{mm}^2$$

Chapter 10
Design to Eurocode 5

10.1 Introduction

Eurocodes are civil and structural engineering design standards, that are published by the European Committee for Standardisation (CEN), for use in all member countries. They are designed to provide a framework for harmonised specifications for use in the construction industry. At present there are nine Eurocodes which are published as European pre-standard, or ENV, to serve as an alternative to the existing national codes in each country. They are as follows:

EC1 Basis of design and actions on structures
EC2 Design of concrete structures
EC3 Design of steel structures
EC4 Design of composite steel and concrete structures
EC5 Design of timber structures
EC6 Design of masonry structures
EC7 Geotechnical design
EC8 Design of structures in seismic regions
EC9 Design of aluminium structures.

The draft edition of Eurocode 5: Part 1.1 for timber structures was published in the UK in late 1994 as DD ENV 1995-1-1 : 1994, in conjunction with a National Application Document (NAD) for the United Kingdom which contains supplementary information to facilitate its use nationally during the ENV period. The publication of the definitive EN EC5 is planned for around 2001 and is expected to be used in parallel with BS 5268 until the year 2005 (overlap period). At the transition, BS 5268 will not be withdrawn but is likely to be termed 'obsolete'. In this state it would not be updated or amended and as such would become less usable with time.

In general, DD ENV 1995-1-1 : 1994 (i.e. EC5) contains all the rules necessary for the design of timber structures, but unlike BS 5268 : Part 2 it does not provide material properties and other necessary design information.

Such information is available in supporting European standards, for example, BS EN 338:1995 which contains the material properties, such as bending and shear strength, etc., for the 15 strength classes of timber (i.e. C14 to D70). Therefore to design with EC5 it is necessary to have several additional reference documents to hand.

10.2   Design philosophy

The design of timber structures is related to Part 1.1 of EC5 and is based on *limit states* design philosophy which means that the overall performance of structures should satisfy two basic requirements. The first is *ultimate limit states* (i.e. *safety*), usually expressed in terms of load-bearing capacity, and the second is *serviceability limit states* (i.e. *deformation and vibration* limits), which refers to the ability of a structural system and its elements to perform satisfactorily in normal use.[1]

In general, it is well understood that the violation of ultimate limit states, i.e. safety criteria, may cause substantial damage and risk to human life, whereas exceeding the serviceability limits, i.e. excessive deformation and/or vibration, rarely leads to such severe consequences. However, while this may suggest that serviceability is relatively unimportant, in practice it is often the key element to ensuring an economical and trouble-free structure.

Generally, it is necessary to check that design criteria for both ultimate limit states and serviceability limit states are satisfied. For checking ultimate limit states, the *characteristic values* of both the loads and the material properties are modified by specified partial safety factors that reflect the reliability of the values that they modify. These factors increase the values of the loads and decrease the values of the material properties. To check serviceability limit states, EC5 requires that both instantaneous and time-dependent (creep) deflections are calculated and, in addition, designers are required to demonstrate that vibration in floors is not excessive.

The *characteristic values* are generally fifth percentile values derived directly from a statistical analysis of laboratory test results. The characteristic strength is the value below which the strength lies in only a small percentages of cases (not more than 5%), and the *characteristic load* is the value above which the load lies in only a small percentage of cases.

Throughout EC5, depending on the character of the individual clauses, a distinction is made between *principles* and *application rules*. Principles are statements, definitions, requirements or analytical models for which no alternative is permitted unless specifically stated (principles are preceded by the letter P), and *application rules* are generally recognised rules that follow the principles and satisfy their requirements.

In EC5, in common with other Eurocodes, decimal points are displayed as commas (so 1.2 is displayed as 1,2).

Table 10.1 Partial safety factors for actions in building structures for persistent and transient design situations (Table 2.3.3.1, EC5)

	Permanent actions γ_G	Variable actions γ_Q	
		One with its characteristic value	Others with their combination value
Normal partial coefficients			
Favourable effect	1.0	0	0
Unfavourable effect	1.35	1.5	1.5
Reduced partial coefficients			
Favourable effect	1.0	0	0
Unfavourable effect	1.2	1.35	1.35

10.3 Actions

Actions is the Eurocode terminology used to define (i) direct loads (forces) applied to the structures, and (ii) indirect forces due to imposed deformations, such as temperature effects or settlement. Actions are covered in Chapter 2 of EC5.

Permanent actions, denoted by the letter G, are all the dead loads acting on the structure, including the self-weight, finishes and fixtures. Variable actions, denoted by the letter Q, are the imposed, wind and snow loads.

Guidance on obtaining the characteristic values of actions is given in EC1 : *Basis of design and actions on structures*. The characteristic values for actions are modified by partial coefficients, γ to take account of safety factors, load combinations, etc.

Therefore the design values of actions, F_d, are obtained by multiplying the characteristic actions for permanent, G_k, and/or variable, Q_k, loads by their relevant partial coefficients, γ_G and γ_Q, as given in Table 2.3.3.1 of EC5, (reproduced here as Table 10.1). Thus:

$$F_d = \sum \gamma_G G_k + \sum \gamma_Q Q_k \tag{10.1}$$

10.4 Material properties

Chapter 3 of ENV 1995-1-1 (EC5 : Part 1.1) deals with the material properties and defines the characteristic strength, stiffness and density values of solid timber sections for strength classes C14 to D70. EC5, unlike BS 5268 : Part 2 : 1996, does not contain the material properties values and, as mentioned earlier, this information can be found in a supporting standard i.e. in Table 1 of BS EN 338 : 1995 (reproduced here as Table 10.2). A comparison of this table with Table 7 of BS 5268 : Part 2 : 1996 (see Table 2.3)

Table 10.2   Characteristic values for structural timber strength classes (Table 1, BS EN 338 : 1995)

Species type		Poplar and conifer species (softwoods)									Deciduous species (hardwoods)					
Strength class		C14	C16	C18	C22	C24	C27	C30	C35	C40	D30	D35	D40	D50	D60	D70
Strength properties (N/mm²)																
Bending	$f_{m,k}$	14	16	18	22	24	27	30	35	40	30	35	40	50	60	70
Tension //	$f_{t,0,k}$	8	10	11	13	14	16	18	21	24	18	21	24	30	36	42
Tension ⊥	$f_{t,90,k}$	0.3	0.3	0.3	0.3	0.4	0.4	0.4	0.4	0.4	0.6	0.6	0.6	0.6	0.7	0.9
Compression //	$f_{c,0,k}$	16	17	18	20	21	22	23	25	26	23	25	26	29	32	34
Compression ⊥	$f_{c,90,k}$	4.3	4.6	4.8	5.1	5.3	5.6	5.7	6.0	6.3	8.0	8.4	8.8	9.7	10.5	13.5
Shear	$f_{v,k}$	1.7	1.8	2.0	2.4	2.5	2.8	3.0	3.4	3.8	3.0	3.4	3.8	4.6	5.3	6.0
Stiffness properties (kN/mm²)																
Mean modulus of elasticity //	$E_{0,mean}$	7	8	9	10	11	12	12	13	14	10	10	11	14	17	20
5%ile modulus of elasticity //	$E_{0,05}$	4.7	5.4	6.0	6.7	7.4	8.0	8.0	8.7	9.4	8.0	8.7	9.4	11.8	14.3	16.8
Mean modulus of elasticity ⊥	$E_{90,mean}$	0.23	0.27	0.30	0.33	0.37	0.40	0.40	0.43	0.47	0.64	0.69	0.75	0.93	1.13	1.33
Mean shear modulus	G_{mean}	0.44	0.50	0.56	0.63	0.69	0.75	0.75	0.81	0.88	0.60	0.65	0.7	0.88	1.06	1.25
Density (kg/m³)																
Characteristic density	ρ_k	290	310	320	340	350	370	380	400	420	530	560	590	650	700	900
Average density	ρ_{mean}	350	370	380	410	420	450	460	480	500	640	670	700	780	840	1080

highlights the difference between the characteristic values and the grade stress values familiar to users of BS 5268. The characteristic values, which are generally fifth percentile values, are considerably higher than the BS 5268 grade values which have been reduced for long-term duration and already include relevant safety factors. In addition, BS EN 338 does not include grade TR26 timber. This is an extra grade which has been included in BS 5268 only.

10.4.1 *Design values*

The characteristic strength values given in Table 1 of BS EN 338 (see Table 10.2) are measured in specimens of certain dimensions conditioned at 20°C and 65% relative humidity, in tests lasting approximately five minutes; and therefore they are applicable for such loading and service conditions. For other loading and service conditions, the characteristic values should be modified using relevant modification factors, which are described later.

Clause 2.2.3.2 of EC5 specifies that the design value, X_d, of a material property is calculated from its characteristic value, X_k, using the following equation:

for strength properties,

$$X_d = \frac{k_{mod} X_k}{\gamma_M}$$
(10.2a)

for stiffness properties,

$$X_d = \frac{X_k}{\gamma_M}$$
(10.2b)

where:

γ_M is the partial safety factor for material property and its values are given in Table 2.3.3.2 of EC5 (reproduced here as Table 10.3)

k_{mod} is the modification factor for service class and duration of load. Modification factor k_{mod} takes into account five different load-duration classes: permanent, long-term, medium-term, short-term and instantaneous, relating to service classes of 1, 2 and 3 as defined in Clauses 3.1.5–3.1.7 of EC5. A summary of Clause 3.1.5 and Tables 3.1.6 and 3.1.7 of EC5 is reproduced here as Table 10.4. For metal components, k_{mod} should be taken as 1, i.e. $k_{mod} = 1.0$.

The characteristic strength values given in Table 1 of BS EN 338, see Table 10.2, are related to a depth in bending and width in tension of 150 mm. For depth in bending or width in tension of solid timber, h, less than 150 mm, EC5 in Clause 3.2.2(5) recommends that the characteristic values for $f_{m,k}$ and $f_{t,0,k}$ may be increased by the factor k_h where:

$$k_h = \text{the lesser of } \left(k_h = \left(\frac{150}{h} \right)^{0.2} \quad \text{and} \quad k_h = 1.3 \right)$$
(10.3)

Table 10.3 Partial coefficients for material properties (Table 2.3.3.2, EC5)

Limit states	γ_M
Ultimate limit states	
Fundamental combinations	
Timber and wood-based materials	1.3
Steel used in joints	1.1
Accidental combinations	1.0
Serviceability limit states	1.0

Table 10.4 Modification factor k_{mod} for service classes and duration of load

Values of k_{mod} for solid and glued laminated timber and plywood (Table 3.1.7, EC5)

Load duration class[a]	Service class[b]		
	1	2	3
Permanent	0.60	0.60	0.50
Long-term	0.70	0.70	0.55
Medium-term	0.80	0.80	0.65
Short-term	0.90	0.90	0.70
Instantaneous	1.10	1.10	0.90

[a] Load duration classes (Table 3.1.6, EC5):
  Permanent refers to order of duration of more than 10 years, e.g. self-weight
  Long-term refers to order of duration of 6 months to 10 years, e.g. storage
  Medium-term refers to order of duration of 1 week to 6 months, e.g. imposed load
  Short-term refers to order of duration of less than 1 week, e.g. wind load
  Instantaneous refers to sudden loading, e.g. accidental load
[b] Service classes (Clause 3.1.5, EC5):
  Service class 1 relates to a typical service condition of 20°C, 65%RH with moisture content $\leq 12\%$
  Service class 2 relates to a typical service condition of 20°C, 85%RH with moisture content $\leq 20\%$
  Service class 3 relates to a service condition leading to a higher moisture content than 20%

10.5 Ultimate limit states

Chapter 5 of EC5 : Part 1.1 sets out the design procedures for members of solid or glued laminated timber with regard to calculation for their strength properties (ultimate limit states) and treats solid and glued laminated members in the same way. Design procedures relevant to flexural and axially loaded members and dowel-type joints are described below.

10.5.1 *Bending*

Members should be designed so that the following conditions are satisfied:

$$k_m \frac{\sigma_{m,y,d}}{f_{m,y,d}} + \frac{\sigma_{m,z,d}}{f_{m,z,d}} \leq 1 \qquad (10.4a)$$

$$\frac{\sigma_{m,y,d}}{f_{m,y,d}} + k_m \frac{\sigma_{m,z,d}}{f_{m,z,d}} \leq 1 \qquad (10.4b)$$

where:

k_m is the modification factor for combined bending stress, detailed in Clause 5.1.6(2) of EC5 as:
for rectangular sections, $k_m = 0.7$
for other cross-sections, $k_m = 1.0$

$\sigma_{m,y,d}$ & $\sigma_{m,z,d}$ are the design bending stresses about the principal axes y–y and z–z, as shown in Fig. 10.1

$$\sigma_{m,y,d} = \frac{M_{y,d}}{W_y}$$

where:
$M_{y,d}$ = design bending moment about y–y axis, and

$W_y = \dfrac{bh^2}{6}$ is the appropriate section modulus

$$\sigma_{m,z,d} = \frac{M_{z,d}}{W_z}$$

where:
$M_{z,d}$ = design bending moment about z–z axis, and

$W_z = \dfrac{hb^2}{6}$ is the appropriate section modulus

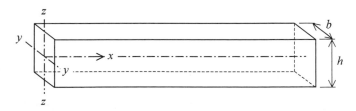

Fig. 10.1 Beam axes.

$f_{m,y,d}$ & $f_{m,z,d}$ are the design bending strengths about y–y and z–z axes

$$f_{m,y/z,d} = \frac{k_{mod} \cdot k_h \cdot k_{crit} \cdot k_{ls} \cdot f_{m,k}}{\gamma_M} \tag{10.5}$$

where:

k_{mod} is the modification factor for load duration and service classes as given in Table 3.1.7 of EC5, see Table 10.4,

k_h is the modification factor for bending depth (Clause 3.2.2),

k_{crit} is the modification factor for reducing bending strength of a beam where there is a possibility of lateral buckling. For a beam which is laterally restrained throughout the length of its compression edge and its ends are prevented from torsional rotation, $k_{crit} = 1.0$. For other conditions, Clause 5.2.2 of EC5 and Clause 6.5 of the UK NAD recommend the following:

(1)   Calculation of the critical bending stress, $\sigma_{m,crit}$, as given by the following equation for a rectangular section of breadth b and depth h:

$$\sigma_{m,crit} = \frac{0.75E_{0,05}b^2}{L_{ef}h} \tag{10.6}$$

The effective length L_{ef} is governed by the degree of restraint against lateral deflection, rotation in plan and twisting, and may be considered as:

$L_{ef} = 0.7L$ for a beam fully restrained against rotation in plan at both ends,

$L_{ef} = 0.85L$ for a beam partially restrained against rotation in plan at both ends or fully restrained at one end,

$L_{ef} = 1.2L$ for a beam partially restrained against twisting at one or both ends.

(2)   Calculation of the relative slenderness for bending as:

$$\lambda_{rel,m} = \sqrt{\left(\frac{f_{m,k}}{\sigma_{m,crit}}\right)} \tag{10.7}$$

(3)   The value of k_{crit} is then determined from:

$k_{crit} = 1$ for $\lambda_{rel,m} \leq 0.75$

$k_{crit} = 1.56 - 0.75\lambda_{rel,m}$ for $0.75 < \lambda_{rel,m} \leq 1.40$

$k_{crit} = 1/\lambda_{rel,m}^2$ for $1.40 < \lambda_{rel,m}$

k_{ls} is the modification factor for load-sharing systems. See Clause 5.4.6 of EC5 for details of the conditions outlined for load sharing systems ($k_{ls} = 1.1$),

$f_{m,k}$ is the characteristic bending strength (Table 1, BS EN 338 : 1995), see Table 10.2,

γ_M is the partial coefficient for material properties (Table 2.3.3.2, EC5), see Table 10.3, and

$E_{0,05}$ is the 5%tile modulus of elasticity (Table 1, BS EN 338 : 1995), see Table 10.2.

10.5.2 *Shear*

Members should be designed so that the following condition is satisfied (Clause 5.1.7, EC5):

$$\tau_d \le f_{V,d} \tag{10.8}$$

where:

τ_d is the design shear stress, and for beams with a rectangular cross-section, is given by:

$$\tau_d = \frac{3 \cdot V_d}{2 \cdot A} \tag{10.9}$$

where V_d is the design shear force (maximum reaction) and A is the cross-sectional area, where $A = bh$. For beams with a notched end $A = bh_e$, see Fig. 10.2.

$f_{V,d}$ is the design shear strength, which is given by:

$$f_{V,d} = \frac{k_{mod} \cdot k_{ls} \cdot k_V \cdot f_{V,k}}{\gamma_M} \tag{10.10}$$

where:

$f_{V,k}$ is the characteristic shear strength (Table 1, BS EN 338 : 1995),

γ_M is the partial coefficient for material properties, (Table 2.3.3.2, EC5),

k_{mod} is the modification factor for load duration and service classes (Table 3.1.7, EC5),

k_{ls} is the modification factor for load-sharing systems ($k_{ls} = 1.1$) (Clause 5.4.6, EC5),

k_V is the modification factor for shear in members with notched ends, see Fig. 10.2, where,
for beams with no notched ends, $k_V = 1.0$
for beams notched at the unloaded side, $k_V = 1.0$
for beams notched at the loaded side,

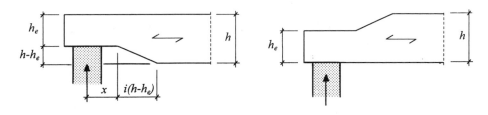

(a) Notch on loaded side   (b) Notch on unloaded side

Fig. 10.2   End-notched beams (DD ENV 1995-1-1 : 1994).

$k_V =$ the lesser of

$$\left(k_V = \frac{k_n\left(1 + \dfrac{1.1 \cdot i^{1.5}}{\sqrt{h}}\right)}{\sqrt{h}\left\{\sqrt{[\alpha(1-\alpha)]} + 0.8\dfrac{x}{h}\sqrt{\left(\dfrac{1}{\alpha} - \alpha^2\right)}\right\}} \quad \text{and} \quad k_V = 1.0 \right)$$

(10.11)

where:

$k_n = 5.0$ for solid timber
$\quad = 6.5$ for glued laminated timber
$h =$ beam depth (mm)
$x =$ distance from line of action to the corner
$\alpha = h_e/h$
$i =$ notch inclination.

10.5.3 *Compression perpendicular to grain (bearing)*

For compression perpendicular to grain the following condition should be satisfied:

$$\sigma_{c,90,d} \leq f_{c,90,d} \tag{10.12}$$

where:

$\sigma_{c,90,d}$ is the design compressive stress perpendicular to grain
$f_{c,90,d}$ is the design compressive strength perpendicular to grain, and is given by:

$$f_{c,90,d} = \frac{k_{mod} \cdot k_{ls} \cdot k_{c,90} \cdot f_{c,90,k}}{\gamma_M} \tag{10.13}$$

where:

$f_{c,90,k}$ is the characteristic compressive strength perpendicular to grain (Table 1, BS EN 338 : 1995),

γ_M is the partial coefficient for material properties (Table 2.3.3.2, EC5),

k_{mod} is the modification factor for load duration and service classes (Table 3.1.7, EC5),

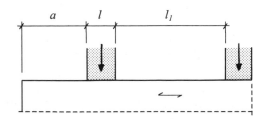

Fig. 10.3 Compression perpendicular to the grain (DD ENV 1995-1-1 : 1994).

k_{ls} is the modification factor for load-sharing systems, i.e. $k_{ls} = 1.1$, (Clause 5.4.6, EC5),

$k_{c,90}$ is the modification factor for bearing length, which takes into account that the load can be increased if the loaded length, l, is short, see Fig. 10.3. The values of $k_{c,90}$ are shown in Table 5.1.5 of EC5 (reproduced here as Table 10.5).

Table 10.5 Values of $k_{c,90}$ (Table 5.1.5, EC5)

| | $l_1 \leq 150$ mm | $l_1 > 150$ mm | |
		$a \geq 100$ mm	$a < 100$ mm
$l \geq 150$ mm	1	1	1
150 mm $> l \geq 15$ mm	1	$1 + (150 - l) / 170$	$1 + a(150 - l) / 17\,000$
15 mm $> l$	1	1.8	$1 + a/125$

10.5.4 Compression or tension parallel to grain

For members subjected to axial compression only, the following condition should be satisfied provided there is no tendency for buckling to occur (Clause 5.1.4, EC5):

$$\sigma_{c,0,d} \leq f_{c,0,d} \tag{10.14}$$

where:

$\sigma_{c,0,d}$ is the design compressive stress parallel to grain, and is given by:

$$\sigma_{c,0,d} = \frac{N_d}{A} \tag{10.15}$$

where N_d is the design axial load and A is the cross-sectional area

$f_{c,0,d}$ is the design compressive strength parallel to grain, and is given by:

$$f_{c,0,d} = \frac{k_{mod} \cdot k_{ls} \cdot f_{c,0,k}}{\gamma_M} \tag{10.16}$$

where:

$f_{c,0,k}$ is the characteristic compressive strength parallel to grain (Table 1, BS EN 338 : 1995),

γ_M is the partial coefficient for material properties (Table 2.3.3.2, EC5),

k_{mod} is the modification factor for load duration and service classes (Table 3.1.7, EC5), and

k_{ls} is the modification factor for load-sharing systems, i.e. $k_{ls} = 1.1$, (Clause 5.4.6, EC5).

For members subjected to *axial tension* a similar condition and procedure apply (Clause 5.1.2, EC5).

10.5.5 *Members subjected to combined bending and axial tension*

For members subjected to combined bending and axial tension, the following conditions should be satisfied (Clause 5.1.9, EC5):

$$\frac{\sigma_{t,0,d}}{f_{t,0,d}} + \frac{\sigma_{m,y,d}}{f_{m,y,d}} + k_m \frac{\sigma_{m,z,d}}{f_{m,z,d}} \leq 1.0 \qquad (10.17a)$$

$$\frac{\sigma_{t,0,d}}{f_{t,0,d}} + k_m \frac{\sigma_{m,y,d}}{f_{m,y,d}} + \frac{\sigma_{m,z,d}}{f_{m,z,d}} \leq 1.0 \qquad (10.17b)$$

where:

$\sigma_{t,0,d}$	is the design tensile stress
$f_{t,0,d}$	is the design tensile strength parallel to grain
$\sigma_{m,y,d}$ & $\sigma_{m,z,d}$	are the respective design bending stresses due to any lateral or eccentric loads (see Section 10.5.1)
$f_{m,y,d}$ & $f_{m,z,d}$	are the design bending strengths
$k_m = 0.7$	for rectangular sections, and
$k_m = 1.0$	for other cross-sections (Clause 5.1.6, EC5).

10.5.6 *Columns subjected to combined bending and axial compression*

For the design of columns subjected to combined bending and axial compression (Clause 5.2.1, EC5), and also for the design of slender columns in general, the following requirements should be considered:

(1) Calculation of the relative slenderness ratios about the y–y and z–z axes of the column as defined by:

$$\lambda_{rel,y} = \sqrt{\left(\frac{f_{c,0,k}}{\sigma_{c,crit,y}}\right)} \qquad (10.18a)$$

$$\lambda_{rel,z} = \sqrt{\left(\frac{f_{c,0,k}}{\sigma_{c,crit,z}}\right)} \qquad (10.18b)$$

where:

$$\sigma_{c,crit,y} = \pi^2 \frac{E_{0,05}}{\lambda_y^2} \qquad (10.19a)$$

$$\sigma_{c,crit,z} = \pi^2 \frac{E_{0,05}}{\lambda_z^2} \qquad (10.19b)$$

where:

$$\lambda_y = \frac{l_{ef,y}}{i_y}, \quad \lambda_z = \frac{l_{ef,z}}{i_z}$$

$E_{0,05}$ is the 5%ile modulus of elasticity parallel to the grain (Table 1, BS EN 338 : 1995).

(2) For both $\lambda_{rel,y} \le 0.5$ and $\lambda_{rel,z} \le 0.5$, the stresses should satisfy the following conditions:

$$\left(\frac{\sigma_{c,0,d}}{f_{c,0,d}}\right)^2 + \frac{\sigma_{m,y,d}}{f_{m,y,d}} + k_m \frac{\sigma_{m,z,d}}{f_{m,z,d}} \le 1.0 \tag{10.20a}$$

$$\left(\frac{\sigma_{c,0,d}}{f_{c,0,d}}\right)^2 + k_m \frac{\sigma_{m,y,d}}{f_{m,y,d}} + \frac{\sigma_{m,z,d}}{f_{m,z,d}} \le 1.0 \tag{10.20b}$$

where:

$\sigma_{c,0,d}$	is the design compressive stress
$f_{c,0,d}$	is the design compressive strength parallel to grain
$\sigma_{m,y,d}$ & $\sigma_{m,z,d}$	are the respective design bending stresses due to any lateral or eccentric loads (see Section 10.5.1)
$f_{m,y,d}$ & $f_{m,z,d}$	are the design bending strengths
$k_m = 0.7$	for rectangular sections, and
$k_m = 1.0$	for other cross-sections (Clause 5.1.6, EC5).

(3) For all other cases, i.e. $\lambda_{rel,y} > 0.5$ and/or $\lambda_{rel,z} > 0.5$, the stresses should satisfy the following conditions:

$$\frac{\sigma_{c,0,d}}{k_{c,z} \cdot f_{c,0,d}} + \frac{\sigma_{m,z,d}}{f_{m,z,d}} + k_m \frac{\sigma_{m,y,d}}{f_{m,y,d}} \le 1.0 \tag{10.21a}$$

$$\frac{\sigma_{c,0,d}}{k_{c,y} \cdot f_{c,0,d}} + k_m \frac{\sigma_{m,z,d}}{f_{m,z,d}} + \frac{\sigma_{m,y,d}}{f_{m,y,d}} \le 1.0 \tag{10.21b}$$

where $k_{c,y}$ and $k_{c,z}$ are modification factors of compression members and are given by:

$$k_{c,y} = \frac{1}{k_y + \sqrt{(k_y^2 - \lambda_{rel,y}^2)}}, \quad \text{(similarly for } k_{c,z}) \tag{10.22a}$$

$$k_y = 0.5(1 + \beta_c(\lambda_{rel,y} - 0.5) + \lambda_{rel,y}^2), \quad \text{(similarly for } k_z) \tag{10.22b}$$

where:

$\beta_c = 0.2$ for solid timber, and
$\beta_c = 0.1$ for glued laminated timber.

10.5.7 Dowel-type fastener joints

EC5, Chapter 6 deals with the ultimate limit states design criteria for joints made with dowel-type fasteners, such as nails, screws, bolts and dowels, but does not give any information regarding connectored-type joints, i.e. toothed plates, split-rings, etc.

EC5 provides formulae, based on Johansen's equations, to determine the load-carrying capacity of timber-to-timber, panel-to-timber and steel-to-timber joints, reflecting all possible failure modes. Load-carrying capacity formulae for timber-to-timber and panel-to-timber joints, with reference to their failure modes, are compiled in Table 10.6, and for steel-to-timber joints in Table 10.7.

Table 10.6 Load-carrying capacity of timber-to-timber and panel-to-timber joints (based on Clause 6.2.1, EC5)

Joints in single shear

Failure modes

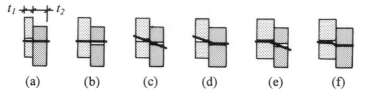

(a)   (b)   (c)   (d)   (e)   (f)

Design load-carrying capacity per dowel, R_d, is the lesser of:

$$R_{d,a} = f_{h,1,d} \cdot t_1 \cdot d \tag{10.23a}$$

$$R_{d,b} = f_{h,1,d} \cdot t_2 \cdot d \cdot \beta \tag{10.23b}$$

$$R_{d,c} = \frac{f_{h,1,d} \cdot t_1 \cdot d}{1+\beta}\left[\sqrt{\left\{\beta + 2\beta^2\left[1 + \frac{t_2}{t_1} + \left(\frac{t_2}{t_1}\right)^2\right] + \beta^3\left(\frac{t_2}{t_1}\right)^2\right\}}\right.$$

$$\left. -\beta\left(1 + \frac{t_2}{t_1}\right)\right] \tag{10.23c}$$

$$R_{d,d} = 1.1\frac{f_{h,1,d} \cdot t_1 \cdot d}{2+\beta}\left\{\sqrt{\left[2\beta(1+\beta) + \frac{4\beta(2+\beta)M_{y,d}}{f_{h,1,d} \cdot t_1^2 \cdot d}\right]} - \beta\right\} \tag{10.23d}$$

$$R_{d,e} = 1.1\frac{f_{h,1,d} \cdot t_2 \cdot d}{1+2\beta}\left\{\sqrt{\left[2\beta^2(1+\beta) + \frac{4\beta(1+2\beta)M_{y,d}}{f_{h,1,d} \cdot t_2^2 \cdot d}\right]} - \beta\right\} \tag{10.23e}$$

$$R_{d,f} = 1.1\sqrt{\left(\frac{2\beta}{1+\beta}\right)}\sqrt{(2M_{y,d} \cdot f_{h,1,d} \cdot d)} \tag{10.23f}$$

Joints in double shear

Failure modes

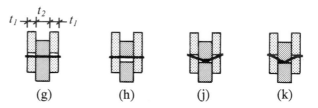

(g)   (h)   (j)   (k)

Design load-carrying capacity per dowel per shear plane, R_d, is the lesser of:

$$R_{d,g} = f_{h,1,d} \cdot t_1 \cdot d \tag{10.23g}$$

$$R_{d,h} = 0.5 f_{h,1,d} \cdot t_2 \cdot d \cdot \beta \tag{10.23h}$$

$$R_{d,j} = 1.1\frac{f_{h,1,d} \cdot t_1 \cdot d}{2+\beta}\left\{\sqrt{\left[2\beta(1+\beta) + \frac{4\beta(2+\beta)M_{y,d}}{f_{h,1,d} \cdot t_1^2 \cdot d}\right]} - \beta\right\} \tag{10.23j}$$

$$R_{d,k} = 1.1\sqrt{\left(\frac{2\beta}{1+\beta}\right)}\sqrt{(2M_{y,d} \cdot f_{h,1,d} \cdot d)} \tag{10.23k}$$

Table 10.7   Load-carrying capacity of steel-to-timber joints (based on Clause 6.2.2, EC5)

Joints in single shear

Failure modes	
	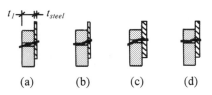
	(a) (b) (c) (d)

Design load-carrying capacity per dowel, R_d, is the lesser of:	(i) Thin steel plate ($t_{steel} \leq 0.5\,d$)	
	$R_{d,a} = 0.4 f_{h,1,d} \cdot t_1 \cdot d$	(10.24a)
	$R_{d,b} = 1.1 \sqrt{(2 M_{y,d} \cdot f_{h,1,d} \cdot d)}$	(10.24b)
	(ii) Thick steel plate ($t_{steel} \geq d$)	
	$R_{d,c} = 1.1 f_{h,1,d} \cdot t_1 \cdot d \left[\sqrt{\left(2 + \dfrac{4 M_{y,d}}{f_{h,1,d} \cdot t_1^2 \cdot d} \right)} - 1 \right]$	(10.24c)
	$R_{d,d} = 1.5 \sqrt{(2 M_{y,d} \cdot f_{h,1,d} \cdot d)}$	(10.24d)

Joints in double shear

Failure modes	
	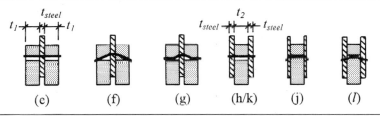
	(e) (f) (g) (h/k) (j) (l)

Design load-carrying capacity per dowel per shear plane R_d, is the lesser of:	(i) Steel plate as the centre member	
	$R_{d,e} = 1.1 f_{h,1,d} \cdot t_1 \cdot d$	(10.24e)
	$R_{d,f} = 1.1 f_{h,1,d} \cdot t_1 \cdot d \left[\sqrt{\left(2 + \dfrac{4 M_{y,d}}{f_{h,1,d} \cdot t_1^2 \cdot d} \right)} - 1 \right]$	(10.24f)
	$R_{d,g} = 1.5 \sqrt{(2 M_{y,d} \cdot f_{h,1,d} \cdot d)}$	(10.24g)
	(ii) Thin steel side plates ($t_{steel} \leq 0.5 d$)	
	$R_{d,h} = 0.5 f_{h,2,d} \cdot t_2 \cdot d$	(10.24h)
	$R_{d,j} = 1.1 \sqrt{(2 M_{y,d} \cdot f_{h,2,d} \cdot d)}$	(10.24j)
	(iii) Thick steel side plates ($t_{steel} \geq d$)	
	$R_{d,k} = 0.5 f_{h,2,d} \cdot t_2 \cdot d$	(10.24k)
	$R_{d,l} = 1.5 \sqrt{(2 M_{y,d} \cdot f_{h,2,d} \cdot d)}$	(10.24l)

As an example, for a two-member timber-to-timber joint there are six possible modes of failure. One or both of the timber members may fail, or the fastener may fail in one or both timber members, see failure modes (a) to (f) in Table 10.6. EC5 therefore provides six different formulae to calculate the failure load for each mode and recommends that the lowest value is taken as the failure load of the connection.

Therefore the design load-carrying capacity, R_d, for a fastener in single shear in a timber-to-timber joint is given as the lesser of $R_{d,a}$ to $R_{d,f}$ (reproduced below from Table 10.6):

$$R_{d,a} = f_{h,1,d} \cdot t_1 \cdot d \tag{10.23a}$$

$$R_{d,b} = f_{h,1,d} \cdot t_2 \cdot d \cdot \beta \tag{10.23b}$$

$$R_{d,c} = \frac{f_{h,1,d} \cdot t_1 \cdot d}{1+\beta}\left[\sqrt{\left\{\beta + 2\beta^2\left[1 + \frac{t_2}{t_1} + \left(\frac{t_2}{t_1}\right)^2\right] + \beta^3\left(\frac{t_2}{t_1}\right)^2\right\}}\right.$$
$$\left. - \beta\left(1 + \frac{t_2}{t_1}\right)\right] \tag{10.23c}$$

$$R_{d,d} = 1.1\frac{f_{h,1,d} \cdot t_1 \cdot d}{2+\beta}\left\{\sqrt{\left[2\beta(1+\beta) + \frac{4\beta(2+\beta)M_{y,d}}{f_{h,1,d} \cdot t_1^2 \cdot d}\right]} - \beta\right\} \tag{10.23d}$$

$$R_{d,e} = 1.1\frac{f_{h,1,d} \cdot t_2 \cdot d}{1+2\beta}\left\{\sqrt{\left[2\beta^2(1+\beta) + \frac{4\beta(1+2\beta)M_{y,d}}{f_{h,1,d} \cdot t_2^2 \cdot d}\right]} - \beta\right\} \tag{10.23e}$$

$$R_{d,f} = 1.1\sqrt{\left(\frac{2\beta}{1+\beta}\right)}\sqrt{(2M_{y,d} \cdot f_{h,1,d} \cdot d)} \tag{10.23f}$$

where:

t_1 & t_2 are timber or board thickness or penetration

$f_{h,1,d}$ & $f_{h,2,d}$ are the design embedding strengths in t_1 and t_2 and are given by:

$$f_{h,1,d} = \frac{k_{mod,1}f_{h,1,k}}{\gamma_M} \quad \text{and} \quad f_{h,2,d} = \frac{k_{mod,2}f_{h,2,k}}{\gamma_M} \tag{10.25}$$

where, $f_{h,1,k}$ and $f_{h,2,k}$ are the characteristic embedding strengths in t_1 and t_2.

The characteristic embedding strength values for different connection types are extracted from Clauses 6.3–6.7 of EC5 and are compiled in Table 10.8. As an example, for a nailed timber-to-timber joint with predrilled holes, $f_{h,k} = 0.082(1 - 0.01d)\rho_k$ in N/mm^2 (Clause 6.3.1.2(1)). Values of the modification factor, k_{mod}, are given in Table 3.1.7 and the values of γ_M are given in Table 2.3.3.2 of EC5.

β $= f_{h,2,d}/f_{h,1,d}$

d is the fastener diameter

$M_{y,d}$ is the fastener yield moment and is given by:

$$M_{y,d} = \frac{M_{y,k}}{\gamma_M} \tag{10.26}$$

where $M_{y,k}$ is the characteristic value for dowel yield moment.

The characteristic yield moment values for different fastener types are extracted from Clauses 6.3–6.7 of EC5 and are compiled in Table 10.9. As an example, for round wire nails, $M_{y,k} = 180d^{2.6}$ in Nmm (Clause 6.3.1.2).

Table 10.8 Values of the characteristic embedding strength, $f_{h,k}$, in N/mm^2 (based on Clauses 6.3–6.7, EC5)

Fastener type	$f_{h,k}$	
	Timber-to-timber Steel-to-timber	Panel-to-timber
Nails ($d \leq 8$ mm) and staples	$0.082\rho_k d^{-0.3}$ no pre-drilling $0.082(1-0.01d)\rho_k$ pre-drilled	$0.11\rho_k d^{-0.3}$ for plywood $30t^{0.6}d^{-0.3}$ for hardboard
Bolts and dowels Parallel to grain, $f_{h,0,k}$ At an angle α to grain	$0.082(1-0.01d)\rho_k$ $f_{h,0,k}/(k_{90}\sin^2\alpha+\cos^2\alpha)$	$0.11(1-0.01d)\rho_k$
Screws ($d \leq 8$ mm) Screws ($d > 8$ mm)	Rules for nails apply Rules for bolts apply	Rules for nails apply Rules for bolts apply

where ρ_k is the joint member characteristics density, given in Table 1 (BS EN 338 : 1995) (in kg/m^3)
   d is the fastener diameter (in mm); for screws, d is the smooth shank diameter
   t is the panel thickness
   $k_{90} = 1.35 + 0.015\,d$, for softwoods
      $= 0.90 + 0.015\,d$, for hardwoods

Table 10.9 Values of fastener yield moment, $M_{y,k}$, in Nmm (based on Clauses 6.3–6.7, EC5)

Fastener type	$M_{y,k}$
Nails – round wire Nails – square	$180d^{2.6}$ $270d^{2.6}$
Bolts and dowels	$0.8f_{u,k}d^3/6$
Screws ($d \leq 8$ mm) Screws ($d > 8$ mm)	Rules for nails apply[a] Rules for bolts apply[a]
Staples	Rules for nails apply

where $f_{u,k}$ is the characteristic tensile strength of the bolt
      (for 4.6 grade steel, $f_{u,k} = 400$ N/mm^2)
   d is the fastener diameter (in mm)
[a] For a screw the effective diameter $d_{ef} = 0.9d$, provided that its root diameter $\geq 0.7d$

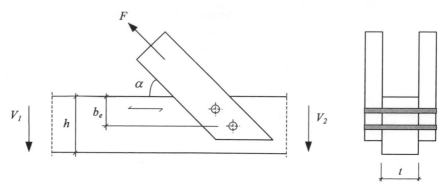

Fig. 10.4 Shear stress in the jointed timber.

EC5, Clause 6.1 recommends that unless a more detailed calculation is made, for the arrangement shown in Fig. 10.4, it should be shown that the following condition is satisfied, provided that $b_e > 0.5\,h$:

$$V_d \leq \tfrac{2}{3} \cdot f_{v,d} \cdot b_e \cdot t \tag{10.27}$$

where:

V_d is the design shear force produced in the member of thickness t by the fasteners at the joint where $(V_1 + V_2 = F \sin \alpha)$

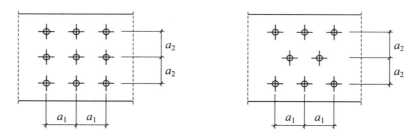

(a) Spacing parallel and perpendicular to grain

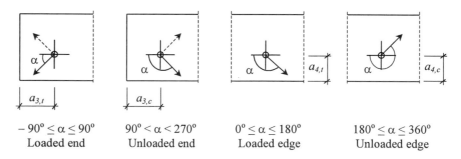

$-90° \leq \alpha \leq 90°$	$90° < \alpha < 270°$	$0° \leq \alpha \leq 180°$	$180° \leq \alpha \leq 360°$
Loaded end	Unloaded end	Loaded edge	Unloaded edge

(b) Edge and end distances

Fig. 10.5 Fastener spacings and distances (DD ENV 1995-1-1 : 1994).

Table 10.10 Minimum nail spacings and distances (based on Table 6.3.1.2, EC5)

Spacings and distances (see Fig. 10.5)	Without pre-drilled holes		Pre-drilled holes
	$\rho_k \leq 420\,\text{kg/m}^3$	$420 < \rho_k < 500\,\text{kg/m}^3$	
Spacing parallel (a_1)	$d < 5\,\text{mm}$: $(5 + 5\lvert\cos\alpha\rvert)d$ $d \geq 5\,\text{mm}$: $(5 + 7\lvert\cos\alpha\rvert)d$	$(7 + 8\lvert\cos\alpha\rvert)d$	$(4 + 3\lvert\cos\alpha\rvert)d$
Spacing perpendicular (a_2)	$5d$	$5d$	$(3 + \lvert\sin\alpha\rvert)d$
Loaded end distance ($a_{3,t}$)	$(10 + 5\lvert\cos\alpha\rvert)d$	$(15 + 5\lvert\cos\alpha\rvert)d$	$(7 + 5\lvert\cos\alpha\rvert)d$
Unloaded end distance ($a_{3,c}$)	$10d$	$15d$	$7d$
Loaded edge distance ($a_{4,t}$)	$(5 + 5\lvert\sin\alpha\rvert)d$	$(7 + 5\lvert\sin\alpha\rvert)d$	$(3 + 4\lvert\sin\alpha\rvert)d$
Unloaded edge distance ($a_{4,c}$)	$5d$	$7d$	$3d$

where d = nail diameter (in mm)

α = angle of force to grain direction

b_e is the distance from the loaded edge to the furthest fastener

$f_{V,d}$ is the design shear strength of timber member.

For nailed connections, EC5, Clause 6.3.1.2 recommends that there should be at least two nails in a joint, that smooth nails should have a pointside penetration of at least $8d$, ($6d$ for improved nails) and nails should be driven into pre-drilled holes in timber with a characteristic density of $500\,kg/m^3$ or more. The recommendations for minimum nail spacings and end distances are given in Table 6.3.1.2 of EC5 (reproduced here as Table 10.10).

For bolted connections, EC5, Clause 6.5.1.2 recommends that for more than six bolts in line with the load direction, the load-carrying capacity of the extra bolts should be reduced by one third, i.e. for n bolts, the effective number, n_{ef}, is $n_{ef} = 6 + 2(n - 6)/3$. The recommendations for minimum bolt and dowel spacings and end distances are given in Tables 6.5.1.2 and 6.6a of EC5 respectively (reproduced here as Table 10.11).

Chapter 6 of EC5 also provides formulae for calculating resistance to axial loading in nailed, screwed and bolted joints, see Clauses 6.3.2 and 6.3.3, 6.7.2 and 6.7.3 and 6.5.2 respectively. These cover withdrawal of the fastener from its head-side (pull-out), pulling through of the fastener head (pull-through), and also combined axial and lateral loading.

Table 10.11   Minimum bolt and dowel spacings and distances (based on Tables 6.5.1.2 and 6.6a, EC5)

Spacings and distances (see Fig. 10.5)	Bolts	Dowels
Spacing parallel (a_1)	$(4 + 3|\cos\alpha|)d$	$(4 + 3|\cos\alpha|)d$
Spacing perpendicular (a_2)	$4d$	$3d$
Loaded end distance ($a_{3,t}$)		
$\quad -90° \leq \alpha \leq 90°$	$7d$ (but $\geq 80\,mm$)	$7d$ (but $\geq 80\,mm$)
Unloaded end distance ($a_{3,c}$)		
$\quad 150° \leq \alpha \leq 210°$	$4d$	$3d$
$\quad 90° < \alpha < 150°$ and $210° < \alpha < 270°$	$(1 + 6|\sin\alpha|)d$ (but $\geq 4d$)	$a_{3,t}6|\sin\alpha|$ (but $\geq 3d$)
Loaded edge distance ($a_{4,t}$)		
$\quad 0° \leq \alpha \leq 180°$	$(2 + 2|\sin\alpha|)d$ (but $\geq 3d$)	$(2 + 2|\sin\alpha|)d$ (but $\geq 3d$)
Unloaded edge distance ($a_{4,c}$)		
$\quad$ all other values of α	$3d$	$3d$

where d = bolt or dowel diameter (in mm)

$\quad\alpha$ = angle of force to grain direction

10.6   Serviceability limit states

Chapter 4 of EC5 : Part 1.1 sets out serviceability requirements with regard to limiting values for deflections and vibrations which are often the governing factors in selection of suitable section sizes. Clause 4.2 of EC5 also provides design procedure for calculation of slip for joints made with dowel-type fasteners, such as nails, screws, bolts and dowels.

10.6.1   *Deflections*

In order to prevent the possibility of damage to surfacing materials, ceilings, partitions and finishes, and to the functional needs as well as any requirements of appearance, Clause 4.3 of EC5 recommends a number of limiting values of deflection for flexural and laterally loaded members. The components of deflection are shown in Fig. 10.6, where the symbols are defined as follows:

u_0 precamber (if applied)
u_1 deflection due to permanent loads
u_2 deflection due to variable loads
u_{net} the net deflection below the straight line joining the supports and is given by:

$$u_{net} = u_1 + u_2 - u_0 \tag{10.28}$$

A summary of EC5 recommendations for limiting deflections is given in Table 10.12.

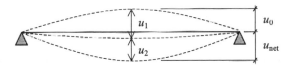

Fig. 10.6   Components of deflection (DD ENV 1995-1-1 : 1994).

Table 10.12   General deflection recommendations (Clause 4.3, EC5)

Components of deflection	Maximum allowable deflection	
	Beams	Cantilevers
Instantaneous deflection due to variable actions, $u_{2,inst}$	$L/300$	$L/150$
Final deflection due to variable actions, $u_{2,fin}$	$L/200$	$L/100$
Final deflection due to all applied actions including precamber, $u_{net,fin}$	$L/200$	$L/100$

where L is the beam span or the length of a cantilever

Table 10.13 Values of k_{def} for solid and glued laminated timber and plywood and joints (based on Table 4.1, EC5)

Material/ Load duration class[a]	Service class[a]		
	1	2	3
Solid and glued laminated timber			
Permanent	0.60	0.80	2.00
Long-term	0.50	0.50	1.50
Medium-term	0.25	0.25	0.75
Short-term	0.00	0.00	0.30
Plywood			
Permanent	0.80	1.00	2.50
Long-term	0.50	0.60	1.80
Medium-term	0.25	0.30	0.90
Short-term	0.00	0.00	0.40

[a] Service and load duration classes are defined in Table 10.4

The UK NAD and EC5 recommend the following procedure for calculation of the instantaneous and final deflections:

(1) The instantaneous deflection of a solid timber member acting alone should be calculated using the appropriate 5%tile modulus of elasticity, $E_{0,05}$ and/or 5%tile shear modulus, $G_{0,05}$, where $G_{0,05} = E_{0,05}/16$.

(2) The instantaneous deflection of a member made of two or more sections fastened together to act as a single member, a member in a load-sharing system, glulam and a composite-section member may be calculated using the mean elastic or shear moduli, i.e. $E_{0,mean}$ and G_{mean}.

(3) The final deflection at the end of the design life of a component, u_{fin}, is calculated as:

$$u_{fin} = u_{inst}(1 + k_{def}) \tag{10.29}$$

where k_{def} is the modification factor which takes into account the increase in deformation with time due to the combined effect of creep and moisture. The values of k_{def} are given in Table 4.1 of EC5 (reproduced here as Table 10.13).

10.6.2 *Vibrations*

EC5, Clause 4.4.1 states that actions which are anticipated to occur frequently should not produce vibrations which might impair the functioning of the structure or cause unacceptable discomfort to the users. Clauses

4.4.2 and 4.4.3 of EC5 provide recommendations and detailed rules for limiting vibrations.

The UK NAD recommends that for residential UK timber floors the requirements of EC5 can be met by ensuring that the total instantaneous deflection of the floor joists under load (i.e. $u_{1,inst} + u_{2,inst}$) does not exceed 14 mm or $L/333$, whichever is the lesser, where L is the span in mm.

10.6.3 *Joint slip*

EC5, Clause 4.2 provides a design procedure for the calculation of slip for joints made with dowel-type fasteners, such as nails, screws, bolts and dowels. In general, the serviceability requirements are to ensure that any deformation caused by slip does not impair the satisfactory functioning of the structure with regard to strength, attached materials and appearance.

The design procedure for calculation of slip for joints made with dowel-type fasteners is summarised as follows:

(1) The instantaneous slip, u_{inst}, in mm, is calculated as:

for serviceability verification, $\quad u_{inst} = \dfrac{F_{ser}}{k_{ser}}$ (10.30a)

for strength verification, $\quad u_{inst} = \dfrac{3 \cdot F_d}{2 \cdot k_{ser}}$ (10.30b)

where:

F_{ser} & F_d are the service load and design load, in N, per fastener per shear plane

k_{ser} is the instantaneous slip modulus, in N/mm, and its values are given in Table 4.2 of EC5 (reproduced here as Table 10.14).

(2) If the characteristic densities of the two joined members ($\rho_{k,1}$ and $\rho_{k,2}$) are different, then ρ_k in formulae given in Table 4.2 of EC5 (see Table 10.14) should be calculated as:

$$\rho_k = \sqrt{(\rho_{k,1} \cdot \rho_{k,2})}$$ (10.31)

(3) The final joint slip, u_{fin}, should be calculated as:

$$u_{fin} = u_{inst}(1 + k_{def})$$ (10.32)

where k_{def} is the modification factor for deformation and is given in Table 4.1 of EC5.

For a joint made from members with different creep properties ($k_{def,1}$ and $k_{def,2}$), the final joint slip should be calculated as:

$$u_{fin} = u_{inst}\sqrt{[(1 + k_{def,1})(1 + k_{def,2})]}$$ (10.33)

Table 10.14 Values of slip moduli k_{ser}, per fastener per shear plane (in N/mm) (Table 4.2, EC5)

Fastener type	k_{ser}
	Timber-to-timber Panel-to-timber Steel-to-timber
Bolts and dowels Screws Nails (pre-drilled)	$\frac{1}{20}\rho_k^{1.5}d$
Nails (no pre-drilling)	$\frac{1}{25}\rho_k^{1.5}d^{0.8}$
Staples	$\frac{1}{60}\rho_k^{1.5}d^{0.8}$

where ρ_k is the joint member characteristics density, given in Table 1, BS EN 338 : 1995 (in kg/m³)
d is the fastener diameter (in mm).

(4)   For joints made with bolts, EC5 permits 1 mm oversize holes which should be added to joint slip values. Thus:

$$u_{inst,bolt} = 1 + \frac{F_{ser}}{k_{ser}} \qquad (10.34)$$

and

$$u_{fin,bolt} = 1 + u_{inst}(1 + k_{def}) \qquad (10.35)$$

10.7   Reference

1.   *Timber Engineering – STEP 1* (1995) Centrum Hout, The Netherlands.

10.8   Bibliography

Timber Engineering – STEP 2 (1955) Centrum Hout, The Netherlands.
Page, A.V. (1993) The new timber design Code : EC5. *The Structural Engineer* **71**, No. 20, 19 October.

10.9   Design examples

Example 10.1   Design of floor joists

A timber floor spanning 3.8 m centre to centre is to be designed using timber joists at 600 mm centres. The floor is subjected to an imposed load of 1.5 kN/m² and carries a dead loading, excluding self-weight, of $0.30\,\text{kN/m}^2$. Carry out design checks to show that a series of $44\,\text{mm} \times 200\,\text{mm}$ deep sawn section timber in strength class C22 under service class 1 is suitable.

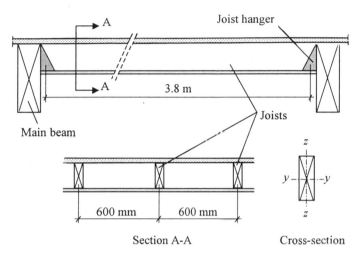

Fig. 10.7 Timber floor joists (Example 10.1).

Definitions

Force, kN	Stress, Nmm^{-2}
Length, m	$N := \text{newton}$
Cross-sectional dimensions, mm	$kN := 10^3 \cdot N$

1. *Geometrical properties*

Breadth of beam section, b $b := 44 \cdot \text{mm}$

Depth of beam section, h $h := 200 \cdot \text{mm}$

Span between support centres, L $L := 3.8 \cdot \text{mm}$

Bearing length, l $l := 75 \cdot \text{mm}$

Joist spacing, Js $Js := 600 \cdot \text{mm}$

Cross-sectional area, A $A := b \cdot h$

$$A = 8.8 \times 10^3 \circ \text{mm}^2$$

Second moment of area about $y\text{–}y$ axis, I_y $I_y := \dfrac{b \cdot h^3}{12}$

$$I_y = 2.93 \times 10^7 \circ \text{mm}^4$$

Section modulus about $y\text{–}y$ axis, W_y $W_y := \dfrac{b \cdot h^2}{6}$

$$W_y = 2.93 \times 10^5 \circ \text{mm}^3$$

2. Timber strength properties

BS EN 338 : 1995, Table 1

Characteristic bending strength	$f_{m.k} := 22 \cdot N \cdot mm^{-2}$
Characteristic shear strength	$f_{V.k} := 2.4 \cdot N \cdot mm^{-2}$
Characteristic compression perpendicular to grain	$f_{c.90.k} := 5.1 \cdot N \cdot mm^{-2}$
Mean modulus of elasticity // to grain	$E_{0.mean} := 10 \cdot kN \cdot mm^{-2}$
5%tile modulus of elasticity // to grain	$E_{0.05} := 6.7 \cdot kN \cdot mm^{-2}$
Mean shear modulus	$G_{mean} := 0.63 \cdot kN \cdot mm^{-2}$
Average density	$\rho_{mean} := 410 \cdot kg \cdot mm^{-3}$

3. Partial safety factors

EC : Part 1.1, Table 2.3.3.1

Permanent actions	$\gamma_G := 1.35$
Variable actions	$\gamma_Q := 1.5$

EC : Part 1.1, Table 2.3.3.2

Material factor for timber	$\gamma_M := 1.3$

4. Actions

Applied dead load, *DL*	$DL := 0.3 \cdot kN \cdot m^{-2}$
Self-weight, *Swt*	$Swt := \rho_{mean} \cdot g \cdot b \cdot h$
	$Swt = 0.04 \circ kN \cdot m^{-1}$
Total characteristic permanent load/m length, G_k	$G_k := DL \cdot Js + Swt$
	$G_k = 0.22 \circ kN \cdot m^{-1}$
Imposed load, *IL*	$IL := 1.5 \circ kN \cdot m^{-2}$
Characteristic variable (imposed) load/m length, Q_k	$Q_k := IL \cdot Js$
	$Q_k = 0.9 \circ kN \cdot m^{-1}$
Design action, F_d	$F_d := \gamma_G \cdot G_k + \gamma_Q \cdot Q_k$
	$F_d = 1.64 \circ kN \cdot m^{-1}$

A. Ultimate limit states

5. Modification factors

Factor for medium-duration loading and service class 1 (k_{mod}, Table 3.1.7)	$K_{mod} := 0.8$
Size factor (k_h, Clause 3.2.2) for $h > 150\,mm$	$k_h := 1.0$

Load sharing applies $k_{ls} := 1.1$
  (k_{ls}, Clause 5.4.6)
Lateral stability
  (k_{crit}, Clause 5.2.2):
Effective length, L_{ef} for full $L_{ef} := 0.7 \cdot L$
  restraint against rotation $L_{ef} = 2.66 \circ \text{m}$

Critical bending stress $\sigma_{m.crit} := \dfrac{0.75 \cdot E_{0.05} \cdot b^2}{L_{ef} \cdot h}$
  (NAD, Clause 6.5)

$$\sigma_{m.crit} = 18.29 \circ \text{N} \cdot \text{mm}^{-2}$$

Relative slenderness for bending $\lambda_{rel.m} := \sqrt{\dfrac{f_{m.k}}{\sigma_{m.crit}}}$

$$\lambda_{rel.m} = 1.1$$

Lateral instability factor $k_{crit} := 1.56 - 0.75 \cdot \lambda_{rel.m}$
$$k_{crit} = 0.74$$

Bearing factor
  ($k_{c,90}$, Clause 5.1.5):
Bearing length, l $l := 75 \cdot \text{mm}$
No overhang at beam ends, a $a := 0 \cdot \text{mm}$
Distance between bearings, l_1 $l_1 := L - l$
$$l_1 = 3.73 \times 10^3 \circ \text{mm}$$

In order to make Mathcad accept the following empirical equation, the indicated units have been added.

Bearing factor $k_{c.90} := 1 + \dfrac{a \cdot (150 \cdot \text{mm} - l)}{17\,000 \cdot \text{mm}^2}$

$$k_{c.90} = 1$$

Factor for shear in members $k_V := 1$ for no notch
  with notched end
  (k_V, Clause 5.1.7.2)

6. *Bending strength*

Design bending moment $M_d := \dfrac{F_d \cdot L^2}{8}$

$$M_d = 2.96 \circ \text{kN} \cdot \text{m}$$

Design bending stress $\sigma_{m.y.d} := \dfrac{M_d}{W_y}$

$$\sigma_{m.y.d} = 10.1 \circ \text{N} \cdot \text{mm}^{-2}$$

Design bending strength $f_{m.y.d} := \dfrac{k_{mod} \cdot k_h \cdot k_{crit} \cdot k_{ls} \cdot f_{m.k}}{\gamma_M}$

$$F_{m.y.d} = 10.98 \circ \text{N} \cdot \text{mm}^{-2}$$
Bending strength satisfactory

7. Shear strength

Design shear load

$$V_d := \frac{F_d \cdot L}{2}$$

$$V_d = 3.12 \circ kN$$

Design shear stress

$$\sigma_{V.d} := \frac{3 \cdot V_d}{2 \cdot A}$$

$$\sigma_{V.d} = 0.53 \circ N \cdot mm^{-2}$$

Design shear strength

$$f_{V.d} := \frac{K_{mod} \cdot k_V \cdot K_{ls} \cdot f_{V.k}}{\gamma_M}$$

$$f_{V.d} = 1.62 \circ N \cdot mm^{-2}$$
Shear strength satisfactory

8. Bearing strength

Design bearing load

$$V_d := \frac{F_d \cdot L}{2}$$

$$V_d = 3.12 \circ kN$$

Design bearing stress

$$\sigma_{c.90.d} := \frac{V_d}{b \cdot l}$$

$$\sigma_{c.90.d} = 0.94 \circ N \cdot mm^{-2}$$

Design bearing strength

$$f_{c.90.d} := \frac{k_{mod} \cdot k_{c.90} \cdot k_{ls} \cdot f_{c.90.k}}{\gamma_M}$$

$$f_{c.90.d} = 3.45 \circ N \cdot mm^{-2}$$
Bearing strength satisfactory

B.   Serviceability limit states

9.   Deflection

Clause 4.3, EC5
For serviceability limit states
(Table 2.3.3.2, EC5)

$$\gamma_M := 1.0$$

Bending moment due to
permanent loads, M_G

$$M_G := \frac{G_k \cdot L^2}{8}$$

$$M_G = 0.39 \circ kN \cdot m$$

Bending moment due to
variable loads, M_Q

$$M_Q := \frac{Q_k \cdot L^2}{8}$$

$$M_Q = 1.62 \circ kN \cdot m$$

(1)   Instantaneous deflection:
For bending and shear due
to permanent action, G_k,
using equations (4.6)
and (4.9)

$$u_{1.inst} := \frac{5 \cdot G_k \cdot L^4}{384 \cdot E_{0.mean} \cdot I_y} + \frac{19.2 \cdot M_G}{(b \cdot h) \cdot E_{0.mean}}$$

$$u_{1.inst} = 2.08 \circ mm$$

For bending and shear due
to variable action, Q_k

$$u_{2.inst} := \frac{5 \cdot Q_k \cdot L^4}{384 \cdot E_{0.mean} \cdot I_y} + \frac{19.2 \cdot M_Q}{(b \cdot h) \cdot E_{0.mean}}$$

$u_{2.inst} = 8.68 \circ \text{mm}$

Maximum allowable
deflection due to variable
action (Clause 4.3, EC5)

$$u_{2.inst.adm} := \frac{L}{300}$$

$u_{2.inst.adm} = 12.67 \circ \text{mm}$

$u_{2.inst} < u_{2.inst.adm}$

Satisfactory

(2) Final deflection due to
variable actions:

Modification factor for
deformation, for
medium-term under
service class 1
(k_{def}, Table 4.1, EC5)

$k_{def} := 0.25$

$u_{2.fin} := u_{2.inst} \cdot (1 + k_{def})$

$u_{2.fin} = 10.86 \circ \text{mm}$

Maximum allowable
deflection
(Clause 4.3, EC5)

$$u_{2.fin.adm} := \frac{L}{200}$$

$u_{2.fin.adm} = 19 \circ \text{mm}$

$u_{2.fin} < u_{2.fin.adm}$

Satisfactory

(3) Final deflection due to all
actions (permanent and
variable):

$u_{net.fin} := (u_{1.inst} + u_{2.inst}) \cdot (1 + k_{def})$

$u_{net.fin} = 13.45 \circ \text{mm}$

Maximum allowable
deflection
(Clause 4.3, EC5)

$$u_{net.fin.adm} := \frac{L}{200}$$

$u_{net.fin.adm} = 19 \circ \text{mm}$

$u_{net.fin} < u_{net.fin.adm}$

Satisfactory

In general, for beams without precamber, it is not necessary to check $u_{2.fin}$ as $u_{net.fin}$ is
always greater.

10. *Vibration*

The UK NAD re: EC5 Clause 4.4.1

Total instantaneous deformations should be less than the lesser of

$$\left(\frac{L}{333} = 11.41 \circ \text{mm} \quad \text{and} \quad 14\,\text{mm} \right)$$

$u_{1.inst} + u_{2.inst} = 10.76 \circ \text{mm}$

Vibration satisfactory

Therefore 44 mm × 200 mm sawn timber sections in strength class C22 are satisfactory

Example 10.2   Design of an eccentrically loaded column

For the design data given below, check that a 100 mm × 250 mm sawn section is adequate as a column if the load is applied 40 mm eccentric about its *y–y* axis. The column is 3.75 m high and has its ends restrained in position but not in direction.

Design data

Timber: C22	Permanent load: 15 kN
Service class: 2	Variable load (medium-term): 17 kN

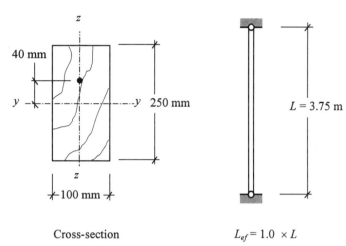

Cross-section

$L_{ef} = 1.0 \times L$

Fig. 10.8   Column details (Example 10.2).

Definitions

Force, kN
Length, m
Cross-sectional dimensions, mm

Stress, Nmm^{-2}
$N := \text{newton}$
$kN := 10^3 \cdot N$

1.   *Geometrical properties*

Column length, L	$L := 3.75 \cdot m$
Effective length, L_{ef}	$L_{ef} := 1.0 \cdot L$
	$L_{ef} = 3.75 \circ m$
Width of section, b	$b := 100 \cdot mm$
Depth of section, d	$d := 250 \cdot mm$
Cross-sectional area, A	$A := b \cdot d$
	$A = 2.5 \times 10^4 \circ mm^2$
Second moment of area about y–y axis	$I_y := \frac{1}{12} \cdot b \cdot d^3$
	$I_y = 1.3 \times 10^8 \circ mm^4$

Section modulus about y–y axis

$$W_y := \frac{b \cdot d^2}{6}$$

$$W_y = 1.04 \times 10^6 \circ \text{mm}^3$$

Radius of gyration about y–y axis

$$i_y := \sqrt{\frac{I_y}{A}}$$

$$i_y = 72.17 \circ \text{mm}$$

Slenderness ratio about y–y axis

$$\lambda_y := \frac{L_{ef}}{i_y}$$

$$\lambda_y = 51.96$$

Second moment of area about z–z axis

$$I_z := \tfrac{1}{12} \cdot d \cdot b^3$$

$$I_z = 2.08 \times 10^7 \circ \text{mm}^4$$

Radius of gyration about z–z axis

$$i_z := \sqrt{\frac{I_z}{A}}$$

$$i_z = 28.87 \circ \text{mm}$$

Slenderness ratio about z–z axis

$$\lambda_z := \frac{L_{ef}}{i_z}$$

$$\lambda_z = 129.9$$

2. Timber strength properties

BS EN 338 : 1995, Table 1

Characteristic bending strength $f_{m.k} := 22 \cdot \text{N} \cdot \text{mm}^{-2}$

Characteristic compression parallel to grain $f_{c.0.k} := 20 \cdot \text{N} \cdot \text{mm}^{-2}$

5%tile modulus of elasticity parallel to grain $E_{0.05} := 6.7 \cdot \text{kN} \cdot \text{mm}^{-2}$

3. Partial safety factors

EC5 : Part 1.1, Table 2.3.3.1

Permanent actions $\gamma_G := 1.35$

Variable actions $\gamma_Q := 1.5$

EC5 : Part 1.1, Table 2.3.3.2

Material for timber $\gamma_M := 1.3$

4. Actions

Characteristic permanent load, G_k $G_k := 15 \cdot \text{kN}$

Characteristic variable load, Q_k $Q_k := 17 \cdot \text{kN}$

Design action, N_d $N_d := \gamma_G \cdot G_k + \gamma_Q \cdot Q_k$

$$N_d = 45.75 \circ \text{kN}$$

Eccentricity, e_y $e_y := 40 \cdot \text{mm}$

Design moment due to eccentricity about y–y axis

$M_d := e_y \cdot N_d$
$M_d = 1.83 \circ \text{kN} \cdot \text{m}$

5. *Modification factors*

Factor for medium-duration loading and service class 1 (k_{mod}, Table 3.1.7)

$k_{mod} := 0.8$

Load sharing does not apply (k_{ls}, Clause 5.4.6)

$k_{ls} := 1.0$

6. *Bending strength*

Design bending moment about y–y axis

$M_d = 1.83 \circ \text{kN} \cdot \text{m}$

Design bending stress about y–y axis

$$\sigma_{m.y.d} := \frac{M_d}{W_y}$$

$\sigma_{m.y.d} = 1.76 \circ \text{N} \cdot \text{mm}^{-2}$

No bending moment about z–z axis, thus

$\sigma_{m.z.d} := 0 \cdot \text{N} \cdot \text{mm}^{-2}$

Design bending strength

$$f_{m.y.d} := \frac{k_{mod} \cdot k_{ls} \cdot f_{m.k}}{\gamma_M}$$

$f_{m.y.d} = 13.54 \circ \text{N} \cdot \text{mm}^{-2}$
$f_{m.z.d} := f_{m.y.d}$ (not required)

7. *Compression strength*

Design compressive stress

$$\sigma_{c.0.d} := \frac{N_d}{A}$$

$\sigma_{c.0.d} = 1.83 \circ \text{N} \cdot \text{mm}^{-2}$

Design compression strength

$$f_{c.0.d} := \frac{k_{mod} \cdot k_{ls} \cdot f_{c.0.k}}{\gamma_M}$$

$f_{c.0.d} = 12.31 \circ \text{N} \cdot \text{mm}^{-2}$

Buckling resistance (Clause 5.2.1, EC5):

Euler critical stresses about y–y and z–z axes

$$\sigma_{c.crit.y} := \frac{\pi^2 \cdot E_{0.05}}{\lambda_y^2}$$

$\sigma_{c.crit.y} = 24.49 \circ \text{N} \cdot \text{mm}^{-2}$

$$\sigma_{c.crit.z} := \frac{\pi^2 \cdot E_{0.05}}{\lambda_z^2}$$

$\sigma_{c.crit.z} = 3.92 \circ \text{N} \cdot \text{mm}^{-2}$

Relative slenderness ratios

$$\lambda_{rel.y} := \sqrt{\frac{f_{c.0.k}}{\sigma_{c.crit.y}}}$$

$$\lambda_{rel.y} = 0.9$$

$$\lambda_{rel.z} := \sqrt{\frac{f_{c.0.k}}{\sigma_{c.crit.z}}}$$

$$\lambda_{rel.z} = 2.26$$

Both relative slenderness ratios are > 0.5, hence conditions in Clause 5.2.1 (4) apply:

For a solid timber section, $\beta_c := 0.2$
For a rectangular section $k_m := 0.7$
  (Clause 5.1.6),
Thus $k_y := 0.5 \cdot [1 + \beta_c \cdot (\lambda_{rel.y} - 0.5) + \lambda_{rel.y}^2]$
     $k_y = 0.95$
and $k_z := 0.5 \cdot [1 + \beta_c \cdot (\lambda_{rel.z} - 0.5) + \lambda_{rel.z}^2]$
     $k_z = 3.23$

Hence $k_{c.y} := \dfrac{1}{k_y + \sqrt{k_y^2 - \lambda_{rel.y}^2}}$

     $k_{c.y} = 0.81$

and $k_{c.z} := \dfrac{1}{k_z + \sqrt{k_z^2 - \lambda_{rel.z}^2}}$

     $k_{c.z} = 0.18$

Check the following conditions: $\dfrac{\sigma_{c.0.d}}{k_{c.z} \cdot f_{c.0.d}} + \dfrac{\sigma_{m.z.d}}{f_{m.z.d}} + k_m \dfrac{\sigma_{m.y.d}}{f_{m.y.d}} = 0.91$

< 1.0 Satisfactory

$$\dfrac{\sigma_{c.0.d}}{k_{c.y} \cdot f_{c.0.d}} + k_m \dfrac{\sigma_{m.z.d}}{f_{m.z.d}} + \dfrac{\sigma_{m.y.d}}{f_{m.y.d}} = 0.31$$

< 1.0 Satisfactory

Therefore a 100 mm × 250 mm sawn section timber in strength class C22 is satisfactory

Example 10.3   Design of a timber-to-timber nailed tension splice joint

A timber-to-timber tension splice joint comprises two 47 mm × 120 mm inner members and two 33 mm × 120 mm side members of strength class C22 timber in service class 2. It is proposed to use 3.35 mm diameter, 65 mm long round wire nails without pre-drilling. The joint is subjected to a permanent load of 2 kN and a medium-term variable load of 3 kN. Determine the required number of nails with a suitable nailing pattern.

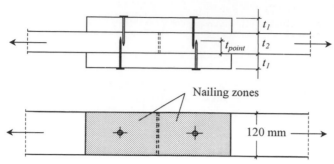

Fig. 10.9(a) Splice joint (Example 10.3).

Definitions

Force, kN	Stress, Nmm^{-2}
Length, m	N := newton
Cross-sectional dimensions, mm	kN := $10^3 \cdot$ N

1. Geometrical properties

Thickness of side members, t_1 $t_1 := 33 \cdot$ mm

Thickness of inner members, t_2 $t_2 := 47 \cdot$ mm

Width of timber members, h $h := 120 \cdot$ mm

Cross-sectional area of side members, A_s $A_s := h \cdot t_1$

$A_s = 3.96 \times 10^3 \circ$ mm^2

Cross-sectional area of inner members, A_{in} $A_{in} := h \cdot t_2$

$A_{in} = 5.64 \times 10^3 \circ$ mm^2

Nail diameter, d $d := 3.35 \cdot$ mm

Nail length, l_{nail} $l_{nail} := 65 \cdot$ mm

Nail pointside penetration, t_{point} $t_{point} := l_{nail} - t_1$

$t_{point} = 32 \circ$ mm

Clause 6.3.1.2, EC5

Minimum allowable pointside penetration $t_{point.adm} := 8 \cdot d$

$t_{point.adm} = 26.8 \circ$ mm

For overlap nailing without pre-drilling $t_2 - t_{point} = 15 \circ$ mm $> 4 \cdot d = 13.4 \circ$ mm

Both pointside penetration and overlap nailing are satisfactory

2. Timber strength properties

BS EN 338 : 1995, Table 1

Characteristic tension parallel to grain $f_{t.0.k} := 13 \cdot$ N $\cdot$ mm^{-2}

Characteristic density $\rho_k := 340 \cdot$ kg $\cdot$ m^{-3}

3. Partial safety factors

EC5 : Part 1.1, Table 2.3.3.1
Permanent actions $\gamma_G := 1.35$
Variable actions $\gamma_Q := 1.5$

EC5 : Part 1.1, Table 2.3.3.2
Material factor for timber $\gamma_{M.timber} := 1.3$
Material factor for steel $\gamma_{M.steel} := 1.1$

4. Actions

Characteristic permanent load, G_k $G_k := 2 \cdot kN$
Characteristic variable load, Q_k $Q_k := 3 \cdot kN$
Design action, N_d $N_d := \gamma_G \cdot G_k + \gamma_Q \cdot Q_k$
$$N_d = 7.2 \circ kN$$

A.   Ultimate limit states

5. Modification factors

Factor for medium-duration $k_{mod} := 0.8$
  loading and service class 1
  (k_{mod}, Table 3.1.7)

Size factor (k_h, Clause 3.2.2) $k_h := 1.3$ and $k_h := \left(\dfrac{150 \ mm}{h} \right)^{0.2}$
  for $h < 150$ mm is the lesser of:
$$k_h = 1.05$$

6. Tension strength of timber

Design tension stress parallel to $\sigma_{t.0.d} := \dfrac{N_d}{2 \cdot A_s}$
  grain in side members
$$\sigma_{t.0.d} = 0.91 \circ N \cdot mm^{-2}$$

Design tension stress parallel to $\sigma_{t.0.d} := \dfrac{N_d}{A_{in}}$
  grain in inner members
$$\sigma_{t.0.d} = 1.28 \circ N \cdot mm^{-2}$$

Design tension strength parallel $f_{t.0.d} := \dfrac{k_{mod} \cdot k_h \cdot f_{t.0.k}}{\gamma_{M.timber}}$
  to grain
$$f_{t.0.d} = 8.37 \circ N \cdot mm^{-2}$$
Tension strength satisfactory

7. Embedding strength of timber

Characteristic embedding strength without pre-drilling

Since the following is an empirical equation (with hidden units in the coefficient part), for Mathcad to produce the correct units, multiply it by the units shown inside the parentheses; or alternatively use the equation without units for ρ_k and d. Thus:

$$f_{h.k} := 0.082 \cdot \rho_k \cdot d^{-0.3} \cdot (\sec^{-2} \cdot m^{2.3} \cdot 1.26 \cdot 10^5)$$
$$f_{h.k} = 19.42 \circ N \cdot mm^{-2}$$

Design embedding strength for headside timber

$$f_{h.1.d} := \frac{k_{mod} \cdot f_{h.k}}{\gamma_{M.timber}}$$
$$f_{h.1.d} = 11.95 \circ N \cdot mm^{-2}$$

for pointside timber

$$f_{h.2.d} := f_{h.1.d}$$
$$f_{h.2.d} = 11.95 \circ N \cdot mm^{-2}$$

8. Yield moment of nails

Yield moment of a nail (using a similar treatment for the units as above)

$$M_{y.k} := 180 \cdot d^{2.6} \cdot (kg \cdot m^{-.6} \cdot \sec^{-2} \cdot 6.31 \cdot 10^4)$$
$$M_{y.k} = 4.17 \times 10^3 \circ N \cdot mm$$

Design yield moment

$$M_{y.d} := \frac{k_{mod} \cdot M_{y.k}}{\gamma_{M.steel}}$$
$$M_{y.d} = 3.03 \times 10^3 \circ N \cdot mm$$

9. Load-carrying capacity

EC5, Clause 6.2.1

For a timber-to-timber joint with nails in single shear, design resistance per shear plane, R_d, is the lesser of $R_{d.a}$ to $R_{d.f}$ (see Table 10.6), where:

t_2 is the pointside penetration length

$$t_2 := t_{point}$$

and the ratio, β

$$\beta := \frac{f_{h.1.d}}{f_{h.2.d}}$$
$$\beta = 1$$

Failure mode (a)

$$R_{d.a} := f_{h.1.d} \cdot t_1 \cdot d$$
$$R_{d.a} = 1.32 \times 10^3 \circ N$$

Failure mode (b)

$$R_{d.b} := f_{h.1.d} \cdot t_2 \cdot d \cdot \beta$$
$$R_{d.b} = 1.28 \times 10^3 \circ N$$

Failure mode (c)

$$R_{d.c} := \frac{f_{h.1.d} \cdot d \cdot t_1}{1 + \beta}$$
$$\cdot \left[\sqrt{\beta + 2 \cdot \beta^2 \cdot \left[1 + \frac{t_2}{t_1} + \left(\frac{t_2}{t_1}\right)^2 \right] + \beta^3 \cdot \left(\frac{t_2}{t_1}\right)^2} - \beta \cdot \left(1 + \frac{t_2}{t_1} \right) \right]$$
$$R_{d.c} = 538.94 \circ N$$

Failure mode (d)

$$R_{d.d} := 1.1 \cdot \frac{f_{h.1.d} \cdot d \cdot t_1}{2 + \beta}$$
$$\cdot \left[\sqrt{2 \cdot \beta \cdot (1 + \beta) + \frac{4 \cdot \beta \cdot (2 + \beta) \cdot M_{y.d}}{f_{h.1.d} \cdot d \cdot t_1^2}} - \beta \right]$$
$$R_{d.d} = 580.68 \circ N$$

Failure mode (e)

$$R_{d.e} := 1.1 \cdot \frac{f_{h.1.d} \cdot d \cdot t_2}{1 + 2 \cdot \beta}$$

$$\cdot \left[\sqrt{2 \cdot \beta^2 \cdot (1 + \beta) + \frac{4 \cdot \beta \cdot (1 + 2 \cdot \beta) \cdot M_{y.d}}{f_{h.1.d} \cdot d \cdot t_2^2}} - \beta \right]$$

$$R_{d.e} = 568.73 \circ \text{N}$$

Failure mode (f)

$$R_{d.f} := 1.1 \cdot \sqrt{\left(\frac{2 \cdot \beta}{1 + \beta} \right)} \cdot \sqrt{(2 \cdot M_{y.d} \cdot f_{h.1.d} \cdot d)}$$

$$R_{d.f} = 542.17 \circ \text{N}$$

Therefore the design resistance per nail is the minimum of:

$$R_d := [R_{d.a}\ R_{d.b}\ R_{d.c}\ R_{d.d}\ R_{d.e}\ R_{d.f}]$$
$$\min(R_d) = 538.94 \circ \text{N}$$

Number of nails per side required is

$$N_{nails} := \frac{N_d}{\min(R_d)}$$

$$N_{nails} = 13.36$$

For a symmetrical nailing pattern adopt 16 nails per side. Thus,

$$N_{nails} := 16$$

10. *Nail spacing*

EC5, Table 6.3.1.2

Nail diameter $\qquad d := 3.35$

Angle to grain $\qquad \alpha := 0$

Minimum spacing parallel
$$a_1 := (5 + 5 \cdot |\cos(\alpha)|) \cdot d$$
$$a_1 = 33.5 \circ \text{mm}$$

Minimum spacing perpendicular
$$a_2 := 5 \cdot d$$
$$a_2 = 16.75 \circ \text{mm}$$

Minimum loaded end distance
$$a_{3.t} := (10 + 5 \cdot |\cos(\alpha)|) \cdot d$$
$$a_{3.t} = 50.25 \circ \text{mm}$$

Minimum unloaded end distance
$$a_{3.c} := 10 \cdot d$$
$$a_{3.c} = 33.5 \circ \text{mm} \quad \text{not required}$$

Minimum loaded edge distance
$$a_{4.t} := (5 + 5 \cdot |\sin(\alpha)|) \cdot d$$
$$a_{4.t} = 16.75 \circ \text{mm} \quad \text{not required}$$

Minimum unloaded edge distance
$$a_{4.c} := 5 \cdot d$$
$$a_{4.c} = 16.75 \circ \text{mm}$$

B. *Serviceability limit states*

11. *Joint slip*
EC5, Clause 4.2

Nail diameter $\qquad d := 3.35 \cdot \text{mm}$

Timber characteristic density $\qquad \rho_k = 340 \circ \text{kg} \cdot \text{m}^{-3}$

Slip modulus (k_{ser}, Table 4.2, EC5) and using a similar treatment for the units of this empirical equation, for nails without pre-drilling

$$k_{ser} := \tfrac{1}{25} \cdot \rho_k^{1.5} \cdot d^{0.8} \cdot (\text{kg}^{-1.5} \cdot \text{m}^{2.7} \cdot \text{N} \cdot 10^{5.4})$$
$$k_{ser} = 659.65 \circ \text{N} \cdot \text{mm}^{-1}$$

Design load for serviceability limit states

$$N_{d.serv} := G_k + Q_k$$
$$N_{d.serv} = 5 \times 10^3 \circ \text{N}$$

Load per nail

$$N_{nail} := \frac{N_{d.serv}}{N_{nails}}$$
$$N_{nail} = 312.5 \circ \text{N}$$

Instantaneous slip

$$u_{inst} := \frac{N_{nail}}{k_{ser}}$$
$$u_{inst} = 0.47 \circ \text{mm}$$

Assuming all nails to slip by the same amount, each inner member will move by 0.47 mm relative to side members.

Therefore the total instantaneous slip is:

$$u_{total} := 2 \cdot u_{inst}$$
$$u_{total} = 0.95 \circ \text{mm}$$

Modification factor for deformation due to creep (k_{def}, Table 4.1, EC5)
for permanent load,
for medium-term variable load,

$$k_{def.G} := 0.6$$
$$k_{def.Q} := 0.25$$

Final slip per nail is calculated as

$$u_{fin} := u_{inst} \cdot \frac{G_k}{N_{d.serv}} \cdot (1 + k_{def.G})$$

$$+ u_{inst} \cdot \frac{Q_k}{N_{d.serv}} \cdot (1 + k_{def.Q})$$
$$u_{fin} = 0.66 \circ \text{mm}$$

Final slip of joint

$$u_{joint.total} := 2 \cdot u_{fin}$$
$$u_{joint.total} = 1.32 \circ \text{mm}$$

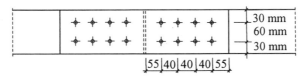

Fig. 10.9(b)   Nail spacings and distances.

Appendix A
Section Sizes for Softwood Timber

In the previous versions of BS 5268 : Part 2 the softwood timber section sizes were specified to BS 4471 : 1978 *Specification for sizes of sawn and processed softwood*, which gave sizes and tolerances for three types of surface finish: sawn, planed and regularised. BS 5268 : Part 2 : 1996 requirements for timber target sizes are those given in BS EN 336 : 1995 *Structural timber. Coniferous and poplar. Sizes. Permissible deviations* and in its National Annex. This standard specifies two tolerance classes: tolerance class 1 (T1) is applicable to sawn surfaces, and tolerance class 2 (T2) applicable to planed timber. Regularised timber can be achieved by specifying T1 for the thickness and T2 for the width.

The commonly available lengths and cross-section sizes are also listed in the National Annex of BS EN 336, and are referred to as target sizes. The *target size* is defined as the desired timber section size (at 20% moisture content) which can be used, without further modification, for design calculations.

Table A1 Basic sawn softwood timber section sizes whose sizes and tolerances comply with BS 4471 : 1978, at 20% moisture content

Thickness (mm)	Width (mm)								
	75	100	125	150	175	200	225	250	300
36	×	×	×	×					
38	×	×	×	×	×	×	×		
44	×	×	×	×	×	×	×	×	×
47	×	×	×	×	×	×	×	×	×
50	×	×	×	×	×	×	×	×	×
63		×	×	×	×	×	×		
75		×	×	×	×	×	×	×	×
100		×		×		×		×	×
150				×		×			×
200						×			
250								×	
300									×

Table A2 Customary target sizes of sawn structural timber (Table NA.2, BS EN 336 : 1995)

Thickness (to tolerance class 1) (mm)	Width (to tolerance class 1) (mm)									
	75	100	125	150	175	200	225	250	275	300
22		×	×	×	×	×	×			
25	×	×	×	×	×	×	×			
38	×	×	×	×	×	×	×			
47	×	×	×	×	×	×	×	×		×
63		×	×	×	×	×	×			
75		×	×	×	×	×	×	×	×	×
100		×		×		×	×	×		×
150				×		×				×
250								×		
300										×

Note 1   Certain sizes may not be obtainable in the customary range of species and grades which are generally available

Note 2   BS EN 336 has a lower limit of 24 mm. However, as thinner material is used in the UK the customary sizes of such material are also listed here

Table A3   Customary lengths of structural timber (Table NA.1, BS EN 336 : 1995)

Length (m)						
1.80	2.10	3.00	4.20	5.10	6.00	7.20
	2.40	3.30	4.50	5.40	6.30	
	2.70	3.60	4.80	5.70	6.60	
		3.90			6.90	

Note   Lengths of 5.70 m and over may not be readily available without finger jointing

In general, the differences between BS 4471 and BS EN 336 are minor and should not present any problems to specifiers and suppliers in the UK.[1] For comparison purposes, in Table A1 the basic sawn softwood timber section sizes whose sizes and tolerances comply with BS 4471 are given. The customary target sizes, whose sizes and tolerances comply with BS EN 336, for sawn strucural timber are given in Table A2. Table A3 gives the range of lengths of sawn softwood structural timber.

Reference

1.   Fewell, A.R. (1997) *Changes to the requirements and supply of timber structural materials*, UKTEG Seminar, I. Struct. Eng., London, February.

Appendix B
Weights of Building Materials

Some typical building material weights, for determination of dead loads, are tabulated in Table B1. The values are based on BS 648 : 1964.

Table B1 Weights of building materials (based on BS 648 : 1964)

Material	Unit mass
Asphalt	
Roofing 2 layers, 19 mm thick	$42\,kg/m^2$
Damp-proofing, 19 mm thick	$41\,kg/m^2$
Bitumen roofing felts	
Mineral surfaced bitumen per layer	$44\,kg/m^2$
Glass fibre	
Slab, per 25 mm thick	$2.5\,kg/m^2$
Gypsum panels and partitions	
Building panels 75 mm thick	$44\,kg/m^2$
Lead	
Sheet, 2.5 mm thick	$30\,kg/m^2$
Linoleum	
3 mm thick	$6\,kg/m^2$
Plaster	
Two coats gypsum, 13 mm thick	$22\,kg/m^2$
Plastic sheeting	
Corrugated	$4.5\,kg/m^2$
Plywood	
per mm thick	$0.7\,kg/m^2$
Rendering or screeding	
Cement : sand (1 : 3), 13 mm thick	$30\,kg/m^2$
Slate tiles	
(depending upon thickness and source)	$24–78\,kg/m^2$
Steel	
Solid (mild)	$7850\,kg/m^3$
Corrugated roofing sheet per mm thick	$10\,kg/m^2$
Tarmacadam	
25 mm thick	$60\,kg/m^2$
Tiling	
Clay, for roof	$70\,kg/m^2$
Timber	
Softwood	$590\,kg/m^3$
Hardwood	$1250\,kg/m^3$
Water	$1000\,kg/m^3$
Woodwool	
Slab, 25 mm thick	$15\,kg/m^2$

Appendix C
Related British Standards for Timber Engineering

BS 5268	**Structural use of timber**
Part 2 : 1996	Code of practice for permissible stress design, materials and workmanship.
Part 3 : 1998	Code of practice for trussed rafter roofs.
Part 4 : 1978	Fire resistance of timber structures.
Part 5 : 1989	Preservation treatments for constructional timber.
Part 6 : 1996	Code of practice for timber frame walls.
Part 7 : 1990	Recommendations for the calculation basis for span tables.
EC5	**DD ENV 1995-1-1 : 1994 Eurocode 5 : Design of timber structures**
	General rules and rules for buildings (together with United Kingdom National Application Document)
BS EN 384 : 1995	Structural timber. Determination of characteristic values of mechanical properties and density.
BS EN 385 : 1995	Finger jointed structural timber. Performance requirements and minimum production requirements.
BS EN 518 : 1995	Structural timber. Grading. Requirements for visual strength grading standards.
BS EN 336 : 1995	Structural timber. Coniferous and poplar. Sizes. Permissible deviations.
BS EN 338 : 1995	Structural timber. Strength classes.
BS EN 519 : 1995	Structural timber. Grading. Requirements for machine strength graded timber and grading machines.

BS EN 301 : 1992   Adhesives, phenolic and aminoplastic, for load-bearing timber structures: classification and performance requirements.

BS EN 386 : 1995   Glued laminated timber. Performance requirements and minimum production requirements.

BS EN 390 : 1995   Glued laminated timber. Sizes. Permissible deviations.

BS EN 20898   Mechanical properties of fasteners.

Index

The Example Worksheets Disks Order Form

All design examples given in this book are produced in the form of worksheets using Mathcad computer software and are available on $3\frac{1}{2}''$ disks to run under Mathcad Version 6, or higher, in either one of its editions: Student, Standard, Plus or Professional. Mathcad runs under Windows Operating System on any IBM compatible Personal Computer.

The worksheet files are labelled examp*m-n*.mcd where m refers to the chapter number and n to the example number in that chapter. For example examp9-2.mcd refers to example 2 in Chapter 9.

It is recommended that the user makes a backup copy of the worksheets. This way he or she is free to experiment with the files. When a worksheet is loaded it should be saved under a new name so that when modified the original disk file is not altered.

Although every care has been taken to ensure the accuracy of the example worksheets, it remains the responsibility of the user to check their results.

Please copy and complete the form at the back of the book, enclosing a cheque or postal order for £20 which includes handling and postage and packing, made payable to A. Kermani, and send to:

A. Kermani, 4 Mid Steil, Glenlockhart, Edinburgh EH10 5XB, UK
email: a.kermani@napier.ac.uk

Structural Timber Design
Mathcad Worksheets

Name: **Title:**

Please indicate your profession

Architect   ☐
Building control officer   ☐
Building engineer   ☐
Civil/structural engineer   ☐
Educator   ☐
Student   ☐

Address:

Postcode:

Date:

Note: There is no warranty with the worksheets.